FIONA: PRINCE EDWARD ISLAND ACCOUNTS OF CANADA'S BIGGEST STORM

Fiona: Prince Edward Island Accounts of Canada's Biggest Storm
ISBN 978-1-7781245-9-4

Edited by Mo Duffy Cobb.
Proofread by Renée Layberry.
Designed by Jordan Beaulieu.
Cover images by Andrew Lewis.
Page 38: From When Things Fall Apart, by Pema Chödrön. Copyright © 1997 by Pema Chödrön. Reprinted by arrangement with The Permissions Company, LLC on behalf of Shambhala Publications, Inc., Boulder, Colorado, shambhala.com.
Printed in Canada by Friesens

All royalties from Fiona: Prince Edward Island Accounts from Canada's Biggest Storm donated to the Island Nature Trust.

Our books may be purchased for promotional, educational, or business use. Please contact your local bookseller or Pownal Street Press at hello@pownalstreetpress.com to purchase.

Pownal Street Press gratefully acknowledges Mi'kma'ki, the ancestral and unceded territory of the Mi'kmaq First Nation on whose land our office is located.

FIONA

PRINCE EDWARD ISLAND ACCOUNTS OF CANADA'S BIGGEST STORM

POWNAL STREET PRESS

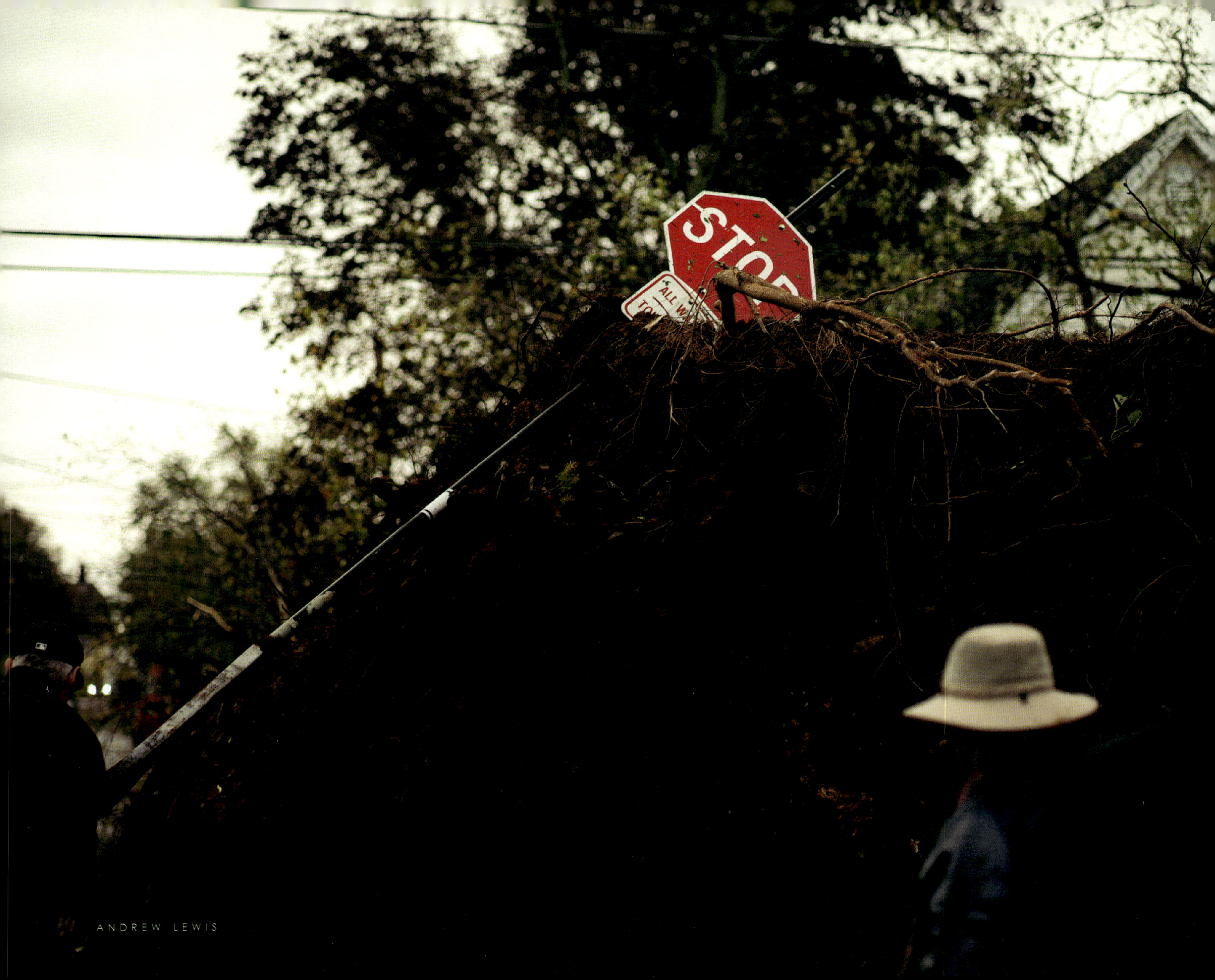
ANDREW LEWIS

TABLE OF CONTENTS

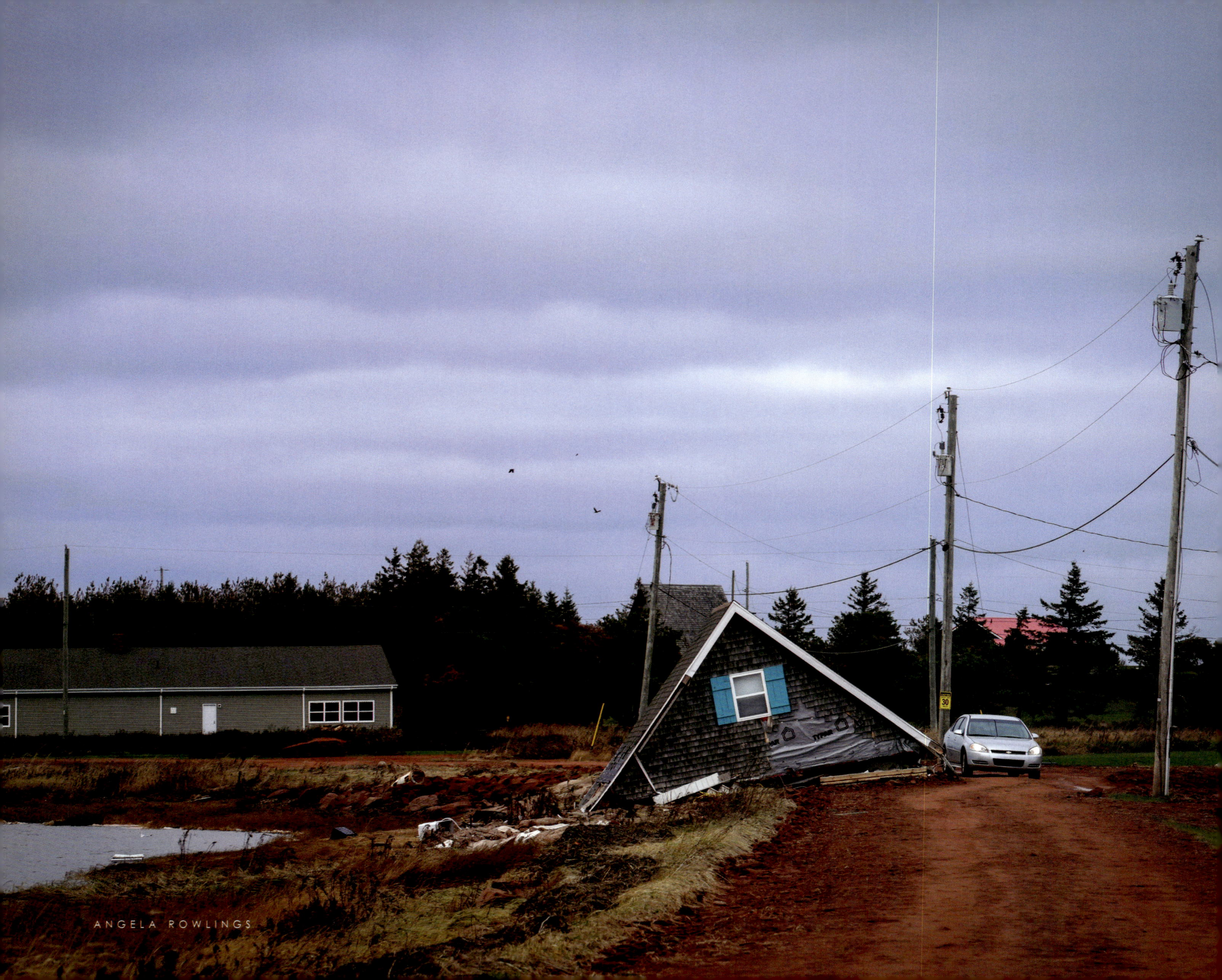

ANGELA ROWLINGS

The Road to Regeneration

IN THE AFTERMATH OF POST-TROPICAL STORM FIONA, our first observations were of the loss of many trees within our cities and towns, our rural communities, and in forests across the province. For many Islanders, trees are an anchoring visual feature in our landscape and in our outdoor experiences.

As information became more available, we understood the widespread nature of the devastation to our forest stands across the island, and many of us were saddened by this. Forests are not all created equal. Because of the history of land use and settlement here on PEI, our forest structures can vary significantly. From what we've seen, single-species or forest plantation stands felt the brunt of Fiona far greater than other forest types such as the mature and structurally diverse Wabanaki-Acadian forest landscapes, which naturally have a greater diversity of species and tree ages.

As an organization dedicated to conservation and land stewardship, we came together to assess the hurricane damage sustained by our 66 protected Natural Areas across the province.

Our Land Stewardship Team monitors our protected Natural Areas year round, and fall is the time when they typically focus on installing signage, training volunteers, planting native trees and shrubs, and organizing beach and illegal dumpsite cleanups, along with doing some minor trail clearing.

In the wake of Fiona, the team, led by Amy Frost-Wicks, pivoted to fully assessing and cleaning up the damage to our protected Natural Areas that contain publicly accessible footpaths. We are acutely aware of the importance of outdoor recreation to both physical and mental well-being. We also believe that, in addition to these benefits, exposure to natural ecosystems helps foster a sense of caring and connection to the land. It was for these reasons that we prioritized making sure our footpaths were safe for public access.

It is in our human nature to want to fix or clean up the forest after destructive events such as post-tropical storm Fiona. However, disturbances like Fiona are a natural component of the Wabanaki-Acadian Forest ecosystem. While the process can be slow from a human perspective, forested ecosystems are in a constant state of growth and change, with young trees growing up to replace those that have fallen. Now is the time to allow ecosystems the space and respect that they need to carry out these processes. While there are actions that we can take in some cases to enhance these processes, such as using the woody debris to build brush piles for wildlife habitat and to speed up decomposition, a healthy forest ecosystem requires minimal human intervention. It is also a perfect opportunity to begin the process of restoring and diversifying forest stands that have a history of human impacts, and which may not be as healthy or resilient as a natural Wabanaki-Acadian forest.

While an intense (and for some, heartbreaking) experience, post-tropical storm Fiona has provided us with an opportunity to observe and adjust our forest management practices to better navigate future severe weather events. We look forward to partnering with other local organizations with the goal of increasing resiliency to events such as this and to other impacts of climate change, along with raising awareness about steps that can and must be taken moving forward. Here's to robust and healthy forests in the future, which support a diversity of species and wildlife, and which all Islanders far and wide can continue to love!

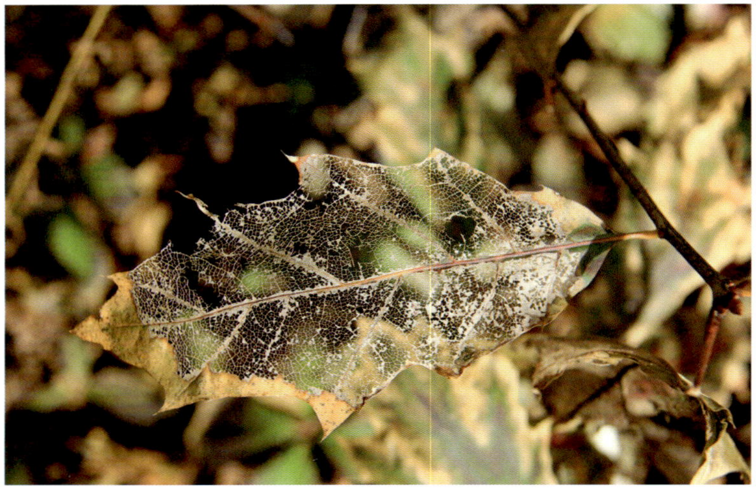

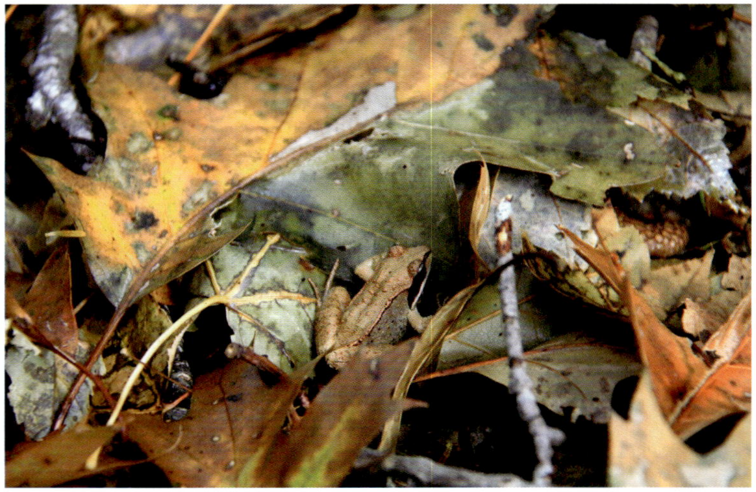

COVERS PRINCE EDWARD ISLAND LIKE THE DEW

FRIDAY, SEPTEMBER 23, 2022

saltwire.com

The Guardian

CHARLOTTETOWN, PRINCE EDWARD ISLAND

A member of the SaltWire Network

WEATHER

Bracing for Fiona

Hurricane could bring 200 mm of rain, 170 km/hr wind gusts to parts of P.E.I.

RAFE WRIGHT
THE GUARDIAN
Rafe.Wright@saltwire.com
@wright542

Hurricane Fiona's course remains fixed on Atlantic Canada and could be one of the biggest storms P.E.I. has experienced in decades.

SaltWire Network's weather specialist Allister Aalders said on Sept. 22 the storm will likely hit sections of the province hard, and residents could see historic amounts of rainfall.

"This does have the potential to be a severe and potentially devastating storm for the region," said Aalders.

On its current course, the storm is likely to make landfall between Guysborough County, N.S., and Cape Breton at a Category 2 strength.

Northeast P.E.I., as well as the Charlottetown area, will see the heaviest rainfall, at 100 to 150 millimetres of rain expected, with 200 millimetres expected in some local areas. Western P.E.I. will likely see anywhere from 50 to 100 millimetres of rain.

Winds across the province are expected to reach gusts of 170 kilometres by Saturday afternoon.

"It's a number that we're not used to using a lot in our projections, it's not an everyday occurrence. Gusts in those range are definitely on the table and likely should be prepared for," said Aalders.

Wind from the storm will start on Friday from a north-northeast direction, but as the centre of the storm passes

SEE BRACING ON A2

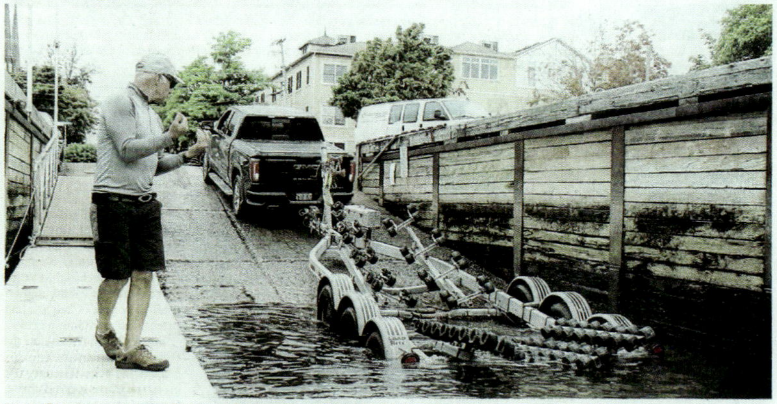

Shane Toombs guides a truck into the Peakes Quay to haul out his boat in Charlottetown on Sept. 22.
RAFE WRIGHT ▪ THE GUARDIAN

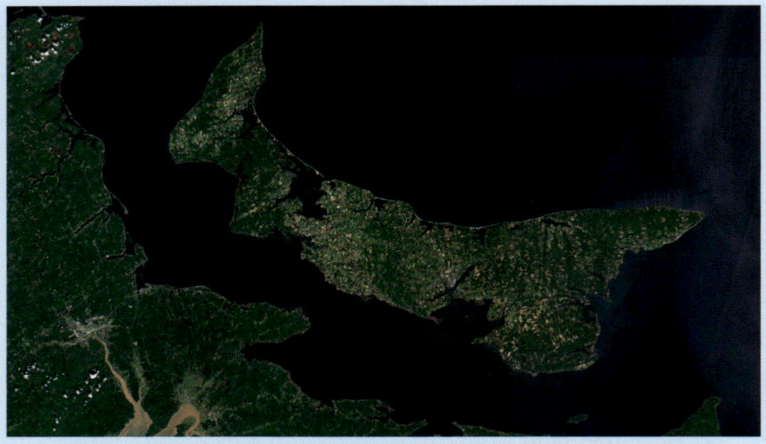

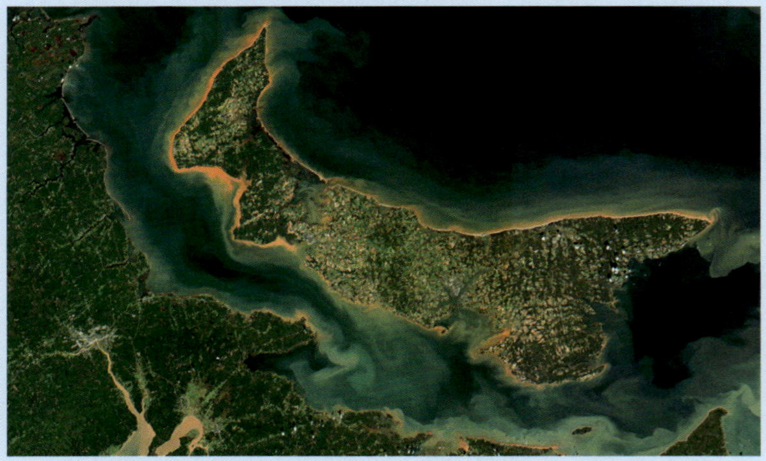

Before and after satellite images show the extent of the destruction on PEI.
(Copernicus Sentinel data 2022)

Foreword

AS THE SUMMER OF 2022 WANED, the Atlantic hurricane season was shaping up to be a dud. Not a single named storm formed in August for the first time in a quarter-century. The last cyclone formed on July 1st, and that storm, named Colin, dissipated just a few days later on July 3rd. July and August were very quiet; September would be anything but.

As we marked the third anniversary of post-tropical storm Dorian on September 7th, the Atlantic basin looked much busier. In addition to Danielle and Earl, two hurricanes tracking well to our east, there was an increasing number of disturbances, known as tropical waves, developing off the coast of West Africa, where many Atlantic hurricanes are born. One of those waves would wreak havoc on the Island as a powerful post-tropical storm in just a few weeks, and Fiona would make Dorian look tame by comparison.

September 23rd was a Friday like no other. A once-minor disturbance flourished over abnormally warm ocean water and soon became a major hurricane bearing down on Atlantic Canada. That night, as Fiona neared the coast of Nova Scotia, it merged with an approaching cold front and transitioned into a post-tropical storm, a classification based on structure, not strength. Unlike a tropical cyclone, a post-tropical storm no longer requires warm ocean water to fuel its intensity; instead, it draws its energy from temperature and pressure differences in the atmosphere. Imagine a powerful nor'easter, laden with tropical moisture, and capable of producing hurricane-strength winds—that is a post-tropical storm.

When Fiona made landfall in eastern Nova Scotia just after 4 AM on Saturday, September 24th, it was as strong as a Category 2 hurricane. As the storm moved over Nova Scotia's Hart Island, the barometric pressure there fell to 931.6 millibars, the lowest pressure ever reported in Canada from a land-falling storm! Fiona was huge. Having shed its symmetrical buzz-saw appearance, Fiona expanded to cover almost all of Atlantic Canada. Heavy rain and a destructive wind lashed all three Island counties well before the storm passed to our east later that day. Because of its immense size, the Island continued to suffer Fiona's wrath long after the storm departed over the Gulf. Unlike Juan and Dorian, this was a long duration event. Fiona was like an ultra-marathon with the intensity of a 100-metre dash. Powerful gusts from 130 km/h to nearly 150 km/h were reported from tip to tip across the Island, damaging roofs, felling limbs and snapping utility poles. The East Point weather station reported seventeen consecutive hours with a peak wind gust of 100 km/h or higher. As 60 mm to well over 100 mm of rain liquefied PEI's renowned red soil, entire trees toppled as their roots could no longer withstand the force of the wind above. This powerful and persistent northerly wind elevated water levels around the Island, producing large waves and a devastating storm surge. Along PEI's north shore, the surge eroded cliffs and calved dunes, stripping away soil, sand and stone. In some of the hardest-hit areas, entire homes and cottages were lifted off their foundations and carried by wind and sea. Sadly, Fiona even toppled Thunder Cove's famous Teacup Rock.

I am a card-carrying weather nerd, yet when I talk about Fiona with family and friends, the peak wind gusts, the central pressure or even the heaviest rainfall measurement rarely come to mind. Meteorologically, these numbers are impressive, but they do not define our experience. With Fiona, the numbers lack nuance; they are better suited for a textbook, not this book. This collection of essays aims to capture the moment and preserve our shared experience of living through a truly historic event for Prince Edward Island.

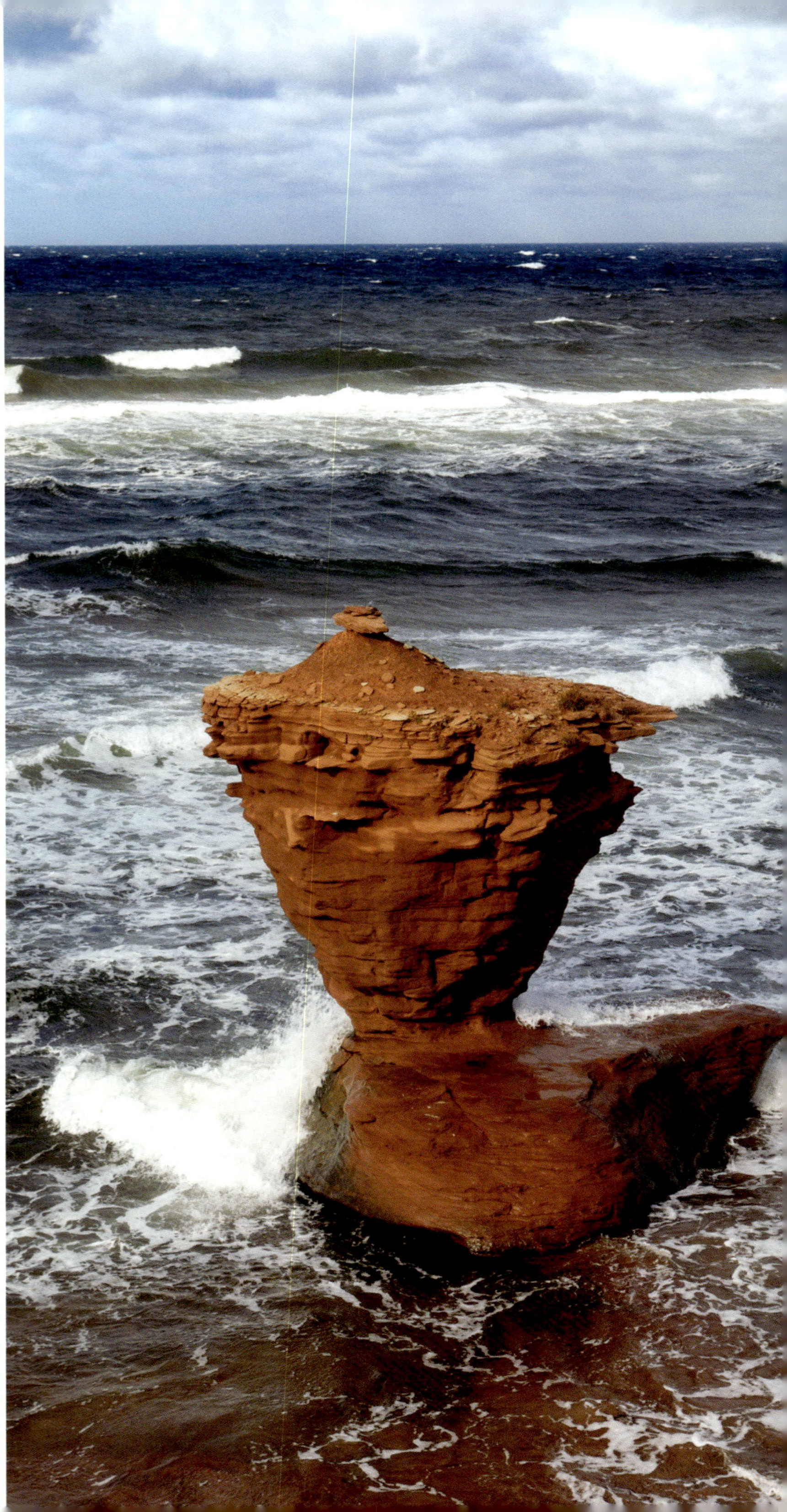

T H E R E ’ S   D I R T   C O M I N G

COVERS PRINCE EDWARD ISLAND LIKE THE DEW

THURSDAY, SEPTEMBER 22, 2022

saltwire.com

The Guardian

CHARLOTTETOWN, PRINCE EDWARD ISLAND

A member of the SaltWire Network

WEATHER

‘We will see an impact’

Hurricane Fiona strengthens to Category 4, continuing north toward Atlantic Canada

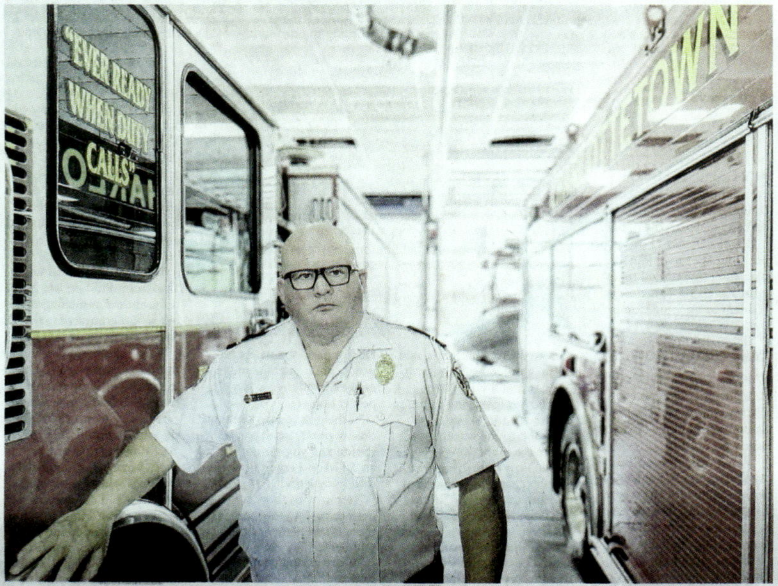

Winston Bryan, fire chief for the City of Charlottetown, says firefighters are preparing for hurricane Fiona and have removed

RAFE WRIGHT
THE GUARDIAN
Rafe.Wright@saltwire.com
@wright542

As hurricane Fiona’s track northward continues to be uncertain, Prince Edward Island residents are being urged to prepare for a weekend of nasty weather.

The storm system strengthened to a Category 4 off the eastern Caribbean on Sept. 21, with winds up to 215 kilometres per hour as of 9 a.m.

At a press conference Sept. 21, put on by the Emergency Measures Organization (EMO) in Charlottetown, Agriculture and Land Minister Darlene Compton said EMO has been paying close attention to the storm over the past few days.

“EMO has been reaching out to all its partners sooner than usual to understand how their planning is going,” said Compton, adding that EMO is working with councillors from municipalities across the province to determine what centres will be open to the public and the precariously housed.

Every municipality decided

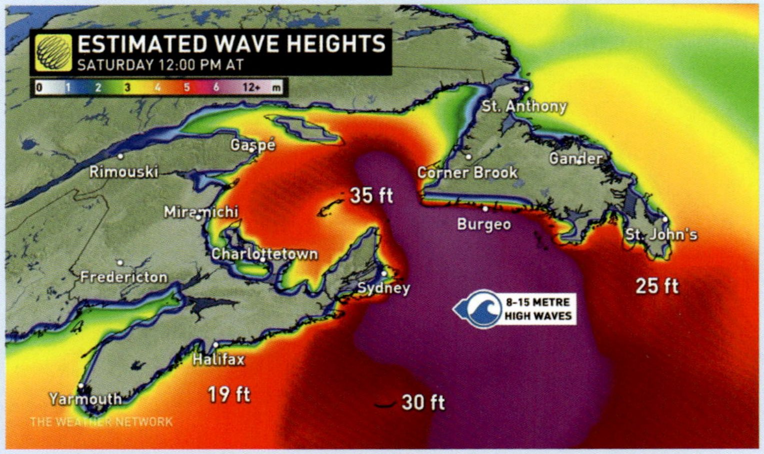

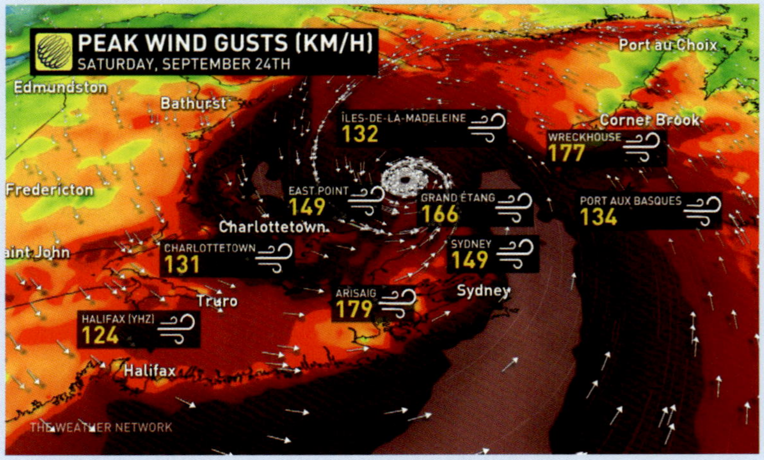

Weather maps predict the extent of the storm throughout Atlantic Canada.
(The Weather Network)

JULIE PELLISSIER-LUSH

The Hurt from Fiona

We heard she was coming, we had to prepare
But how could we know the force of her air?

Through the night we heard the wind
Trees snapping, power out
House by house, road by road
Nothing left untouched

As day broke, we saw the destruction
Homes covered in debris
Some were lucky, many were not
From tip to tip we mourned for our trees

Services gone, we felt so alone
Connection lost
Power lines down. We were all without
Confident were those with generators, but gas hard to find
Even if we could drive out of our own front yards, where could we go?

Flashbacks to our days of lockdown. Cancellations, one by one
Truth and Reconciliation Day and Treaty Day gatherings too
"Just stay put." "Let the linesmen work!"
Mental health falling as the days went by
We knew we lost more than just our electricity with this storm

They say we're only given what we can handle, but
How much more can we all take?
We are strong and resilient, that is true
But Mother Nature will always

hold
the
reins

STELLA SHEPARD

The Spirit of the Trees

IN EARLY CHILDHOOD, I developed a personal relationship with trees, having grown up in rural Prince Edward Island. The woods near my childhood home were a year-round playground. It was a fun place to enjoy tree climbing and view the world like a crow, perched on a treetop. My siblings and I created forts in the woods and built amazing treehouses where we would get lost in our imagination and play games.

Trees are good friends; they are old souls of the universe and have purpose. They absorb carbon dioxide and release oxygen to support life on earth. Trees are healers with a lifelong history of providing food, medicines and shelter for humans and wildlife. They have witnessed world history for millions of years, since the birth of Mother Earth.

Just steps from the farmhouse I live in, a grove of sugar maple trees was planted more than a hundred years ago. I built a medicine wheel in the centre of the grove under the tree canopy. An aura of peacefulness always encompassed the sacred space as I burned sage, beat out a rhythm on my hand drum, and offered prayers of gratitude to the Creator.

An Indigenous Elder once told me that trees have a spirit, and they weep when they are dying. The night Fiona hit the Island, I watched in horror—by flashlight—when those giant sugar maples bent low to the ground. I heard the tortured maples screeching in terror as the force of the winds ripped those mighty giants from the womb of Mother Earth.

The next day, I reluctantly ventured outside. The towering sugar maple trees that protected the medicine wheel were stretched out on the ground like willowy corpses. One of the trees landed inches from the farmhouse, its branches brushing the front door. But the 150-year-old farmhouse had been spared any damage. In deep sorrow, I laid my head on a massive tree trunk and sobbed my heart out, mourning openly the death of the trees.

I snapped the tip from the top of a tree, poked it into a pot of soil and placed it in the warmth of the kitchen. The leaves dropped but the tip of the tree took root and is being reborn.

The rooted tip that once greeted the sky is now flourishing and continues to grow stronger every day, which is amazing to witness. Hurricane Fiona did not kill its spirit. The tiny tree tip is a life lesson of resilience and hope in the aftermath of destruction and grief.

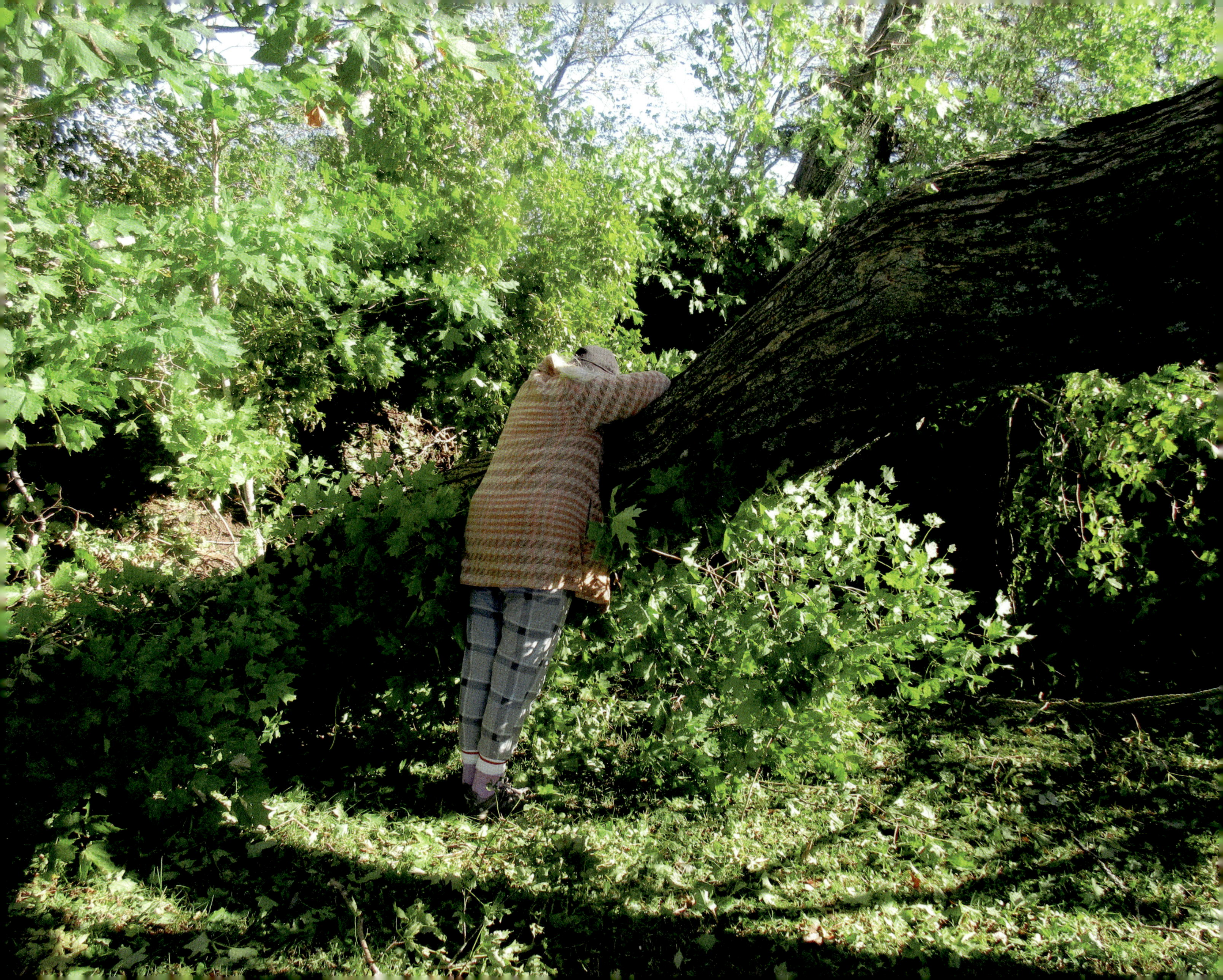

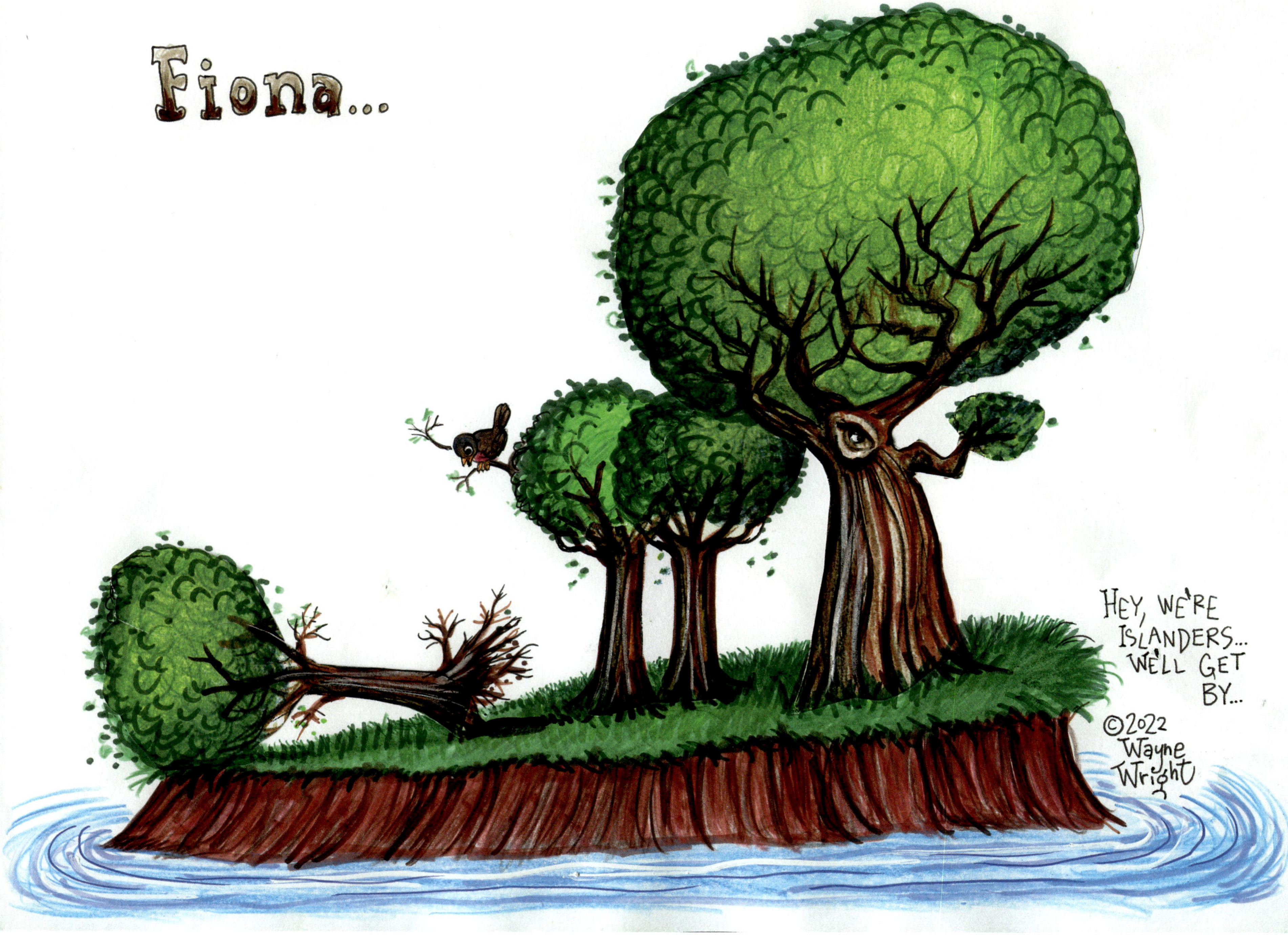
Fiona...
HEY, WE'RE ISLANDERS... WE'LL GET BY...
©2022 Wayne Wright

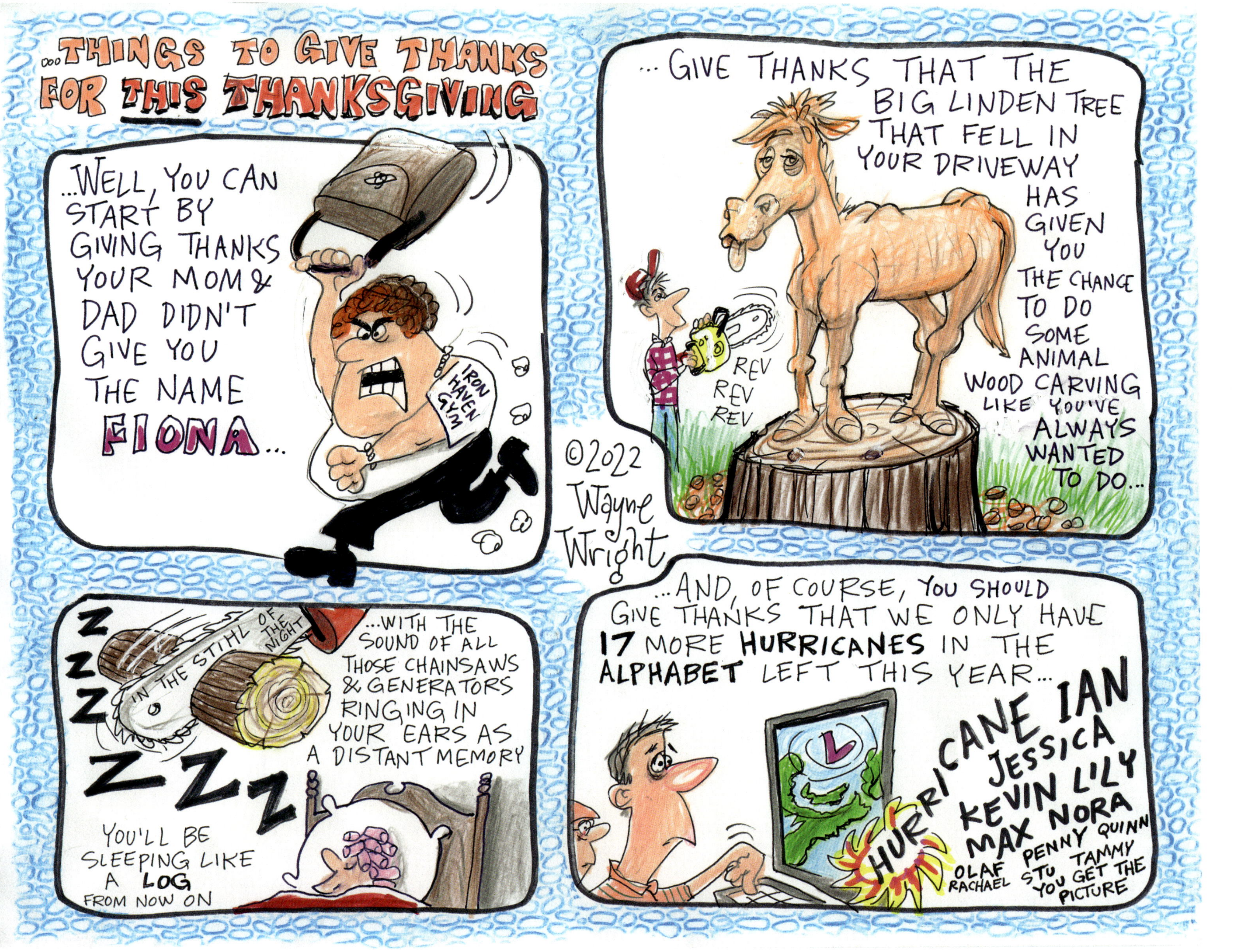
...THINGS TO GIVE THANKS FOR THIS THANKSGIVING
...WELL, YOU CAN START BY GIVING THANKS YOUR MOM & DAD DIDN'T GIVE YOU THE NAME FIONA...
IRON HAVEN GYM
...GIVE THANKS THAT THE BIG LINDEN TREE THAT FELL IN YOUR DRIVEWAY HAS GIVEN YOU THE CHANCE TO DO SOME ANIMAL WOOD CARVING LIKE YOU'VE ALWAYS WANTED TO DO...
REV REV REV
©2022 Wayne Wright
IN THE STIHL OF THE NIGHT
...WITH THE SOUND OF ALL THOSE CHAINSAWS & GENERATORS RINGING IN YOUR EARS AS A DISTANT MEMORY
Z Z Z Z Z Z
YOU'LL BE SLEEPING LIKE A LOG FROM NOW ON
...AND, OF COURSE, YOU SHOULD GIVE THANKS THAT WE ONLY HAVE 17 MORE HURRICANES IN THE ALPHABET LEFT THIS YEAR...
HURRICANE IAN
JESSICA
KEVIN LILY
MAX NORA
OLAF PENNY QUINN
RACHAEL STU TAMMY
YOU GET THE PICTURE

LORRAINE CLEMENTS

Fiona the Fierce

AT 3:00 AM, I WAS WRENCHED FROM SLEEP. I wondered why there were so many chairs scraping on our hardwood floor, and said, "What a noisy dream!" as I bolted upright, not appreciating the violent wake-up. My heart pounded and it deafened me. I shook my husband, Fred, awake and asked if he heard that noise. It sounded like several planes flying low. "What the blazes is going on?" I asked Fred. "Did Putin attack us?"

To say I was scared would be mild. I wasn't scared, I was terrified. In fact, I was too scared even to go into the adjoining bathroom for my nightly call of nature. My husband rolled over on his good ear and started to snore. His deafness was an asset for him the night Fiona assaulted Prince Edward Island. I woke Fred up again, saying, "Honey, it's time we go to the basement." The house shuddered once again, and I could hear crackling and a huge boom. We raced for the stairs to go to the den, one level down.

We sat in the dark, as the power was out. I thought the storm would blow over by morning, but by 4 AM, hell's door was starting to open wider—and I imagined the roof lifting right off the house. As I faced the fact that we might die, I started to pray silently until I heard another crack and three big snaps. Fred lay on the couch, turning over to snore once in a while. I looked out the window towards the darkening sky. The wind in the dark seemed to twist the trees, bending them down to the ground. By 4:30 AM, I decided to weather the storm and put in my earbuds to drown out the sound of the ferocious wind. To my dismay, even the music on my iPod couldn't drown out the wind. The only way I could soothe my soul was to thank God we came into town from the cottage in Rustico. As the house shuddered above us, Fred snored in time with the wind.

By 5 AM, I was thinking, "God, please help us and our families, especially Michael who is out tonight with the North River Fire Department."

If anyone had told me prior to the storm that our street would look like a bomb hit it, I would have laughed it off. You just had to see it to believe it. It was horrifying to see the uprooted trees, taking yards and yards of dirt and roots up with them as they hit the pavement. Yet, no one was hurt.

My sister arrived at our house, and I was shocked when she told me her husband, Wendell Toombs, had passed away of natural causes. I had no words to describe the feeling of loss; the trauma of the storm had left me numb and

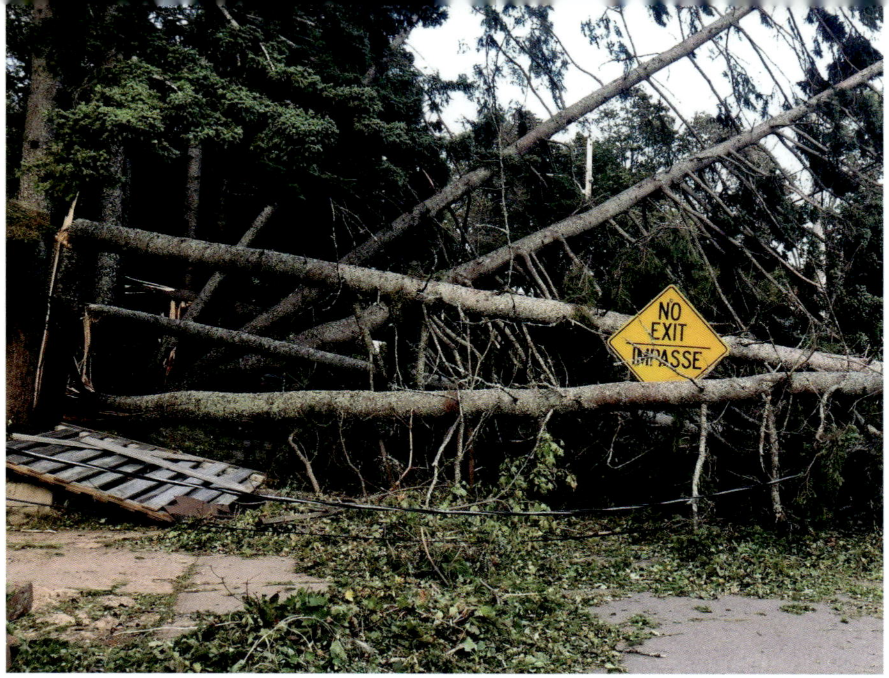

speechless. We weaved our way in and out of debris to reach the funeral home. We ought to have known it wouldn't be open, but grief often creates gaps in the logical thinking process. As our phones were not active, we really had no idea what to do, and only days later contact was made. It was over a week later, on October 5th, that we visited the funeral home to make arrangements for the funeral.

Fred and I soon developed sore throats and became ill with the start of a cold. I blamed the change in the cool air, but little did we know approaching the Thanksgiving weekend that we had contracted Covid, testing positive on our home tests. What a predicament! Saturday and Sunday the Covid testing clinics were closed due to the holiday, so the first availability for PCR testing was Tuesday. We both tested positive and continued to isolate.

As far back as I can remember, Mother Nature has been kind to us. Fiona was a new threat in our world of rapidly rising temperatures and climate change. Warnings after Summer Juan, White Juan and Hurricane Dorian helped prepare Islanders, but the impact of Fiona was almost like the damage from both Juan and Dorian put together. Are post-tropical cyclones like "Fiona the Fierce" here to stay?

CARIN MAKUZ

Because of Fiona

SEPT. 22, 10:00 AM — At the beach christening my new wellies in salt water, singing to the seals, waving to Nova Scotia, dark skies and thunder across the strait. There's more and more talk about a storm on its way. I chat with a woman who spends most of the year at her cottage just the other side of the marram grass. She's heading back to her house to "batten down the hatches," as she says with a chuckle. We keep it light, neither of us wanting to make a big deal of the forecast, as if it's merely an inconvenience. But there's tension too. "See you after," we say.

SEPT. 23, 6:00 PM — It's suddenly much darker, and raining. The wind, strangely still for the last few hours, has picked up. I'm walking around the house, eating my storm chips early.

SEPT. 23, 9:30 PM — Feels like the game has started. Holy moly.

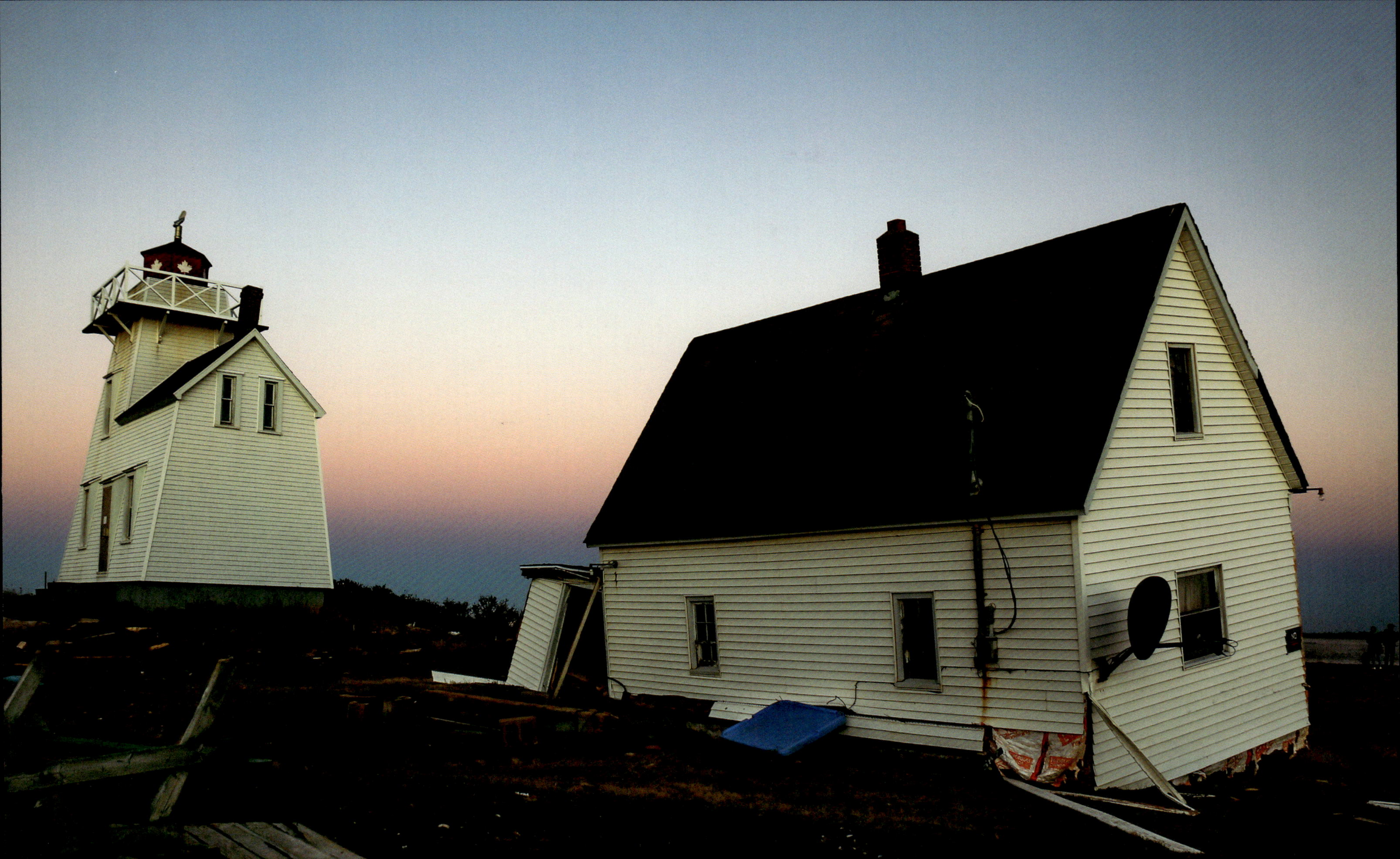

SHIRLEY CALLAGHAN

Night Howl

THE FIRST EARLY WINDS SEEMED INNOCUOUS, but then came the gales and howls of Fiona, the worst post-tropical storm in Canadian history. With the sound of a runaway train and the ferocity of a cheetah, Fiona slashed at everything in her path, intent on the demolition of everything she touched. I was prepared for a storm, but not for that horrifying sound that kept me awake the night she began her chilling attack on the Island. Charlottetown was under siege.

In the morning, my husband and I saw the lovely canopy of trees on our street ripped apart and the woods in the lot behind us uprooted and smashed like matchsticks. There was no power, no internet, and we had no cell phone, leaving us cut off completely. We thought of previous hurricanes—Dorian, then Juan—which were both so devastating in their own right. We were in Hampshire during Juan, and until now we thought it could never be worse than that.

September 24th was my birthday, and a day of justifiable fear. My husband and I sat quietly, thinking of family and friends, and not knowing if we were quite safe yet; we worried especially for the well-being of family at their cottage in South Rustico, near to the surge of the Gulf of St. Lawrence. A neighbour braved the winds to come and see if we were okay. It was like we were in a frightful movie, trapped in our little bungalow.

The next day, there was movement on our street with excavators, payloaders and trucks in action clearing all the trees, branches and leaves which had blocked off Prince Charles Drive. Now, at least ambulances and fire trucks could get by, which was a comfort to us seniors. Up and down our streets was evidence of the force of such a storm.

One neighbour, whose household trees had been completely uprooted, mentioned Haiti and how we thought such a tragedy could never happen here at home. The pride we had in our trees and street was hurt, yet we soon learned of homes and cottages with more severe water and roof damage, farms with their animals and crops in jeopardy, and fishers with their livelihoods threatened. Our lives had been dramatically altered and our sense of indestructibility was shattered, but considering the scope of the damage, we fared out okay.

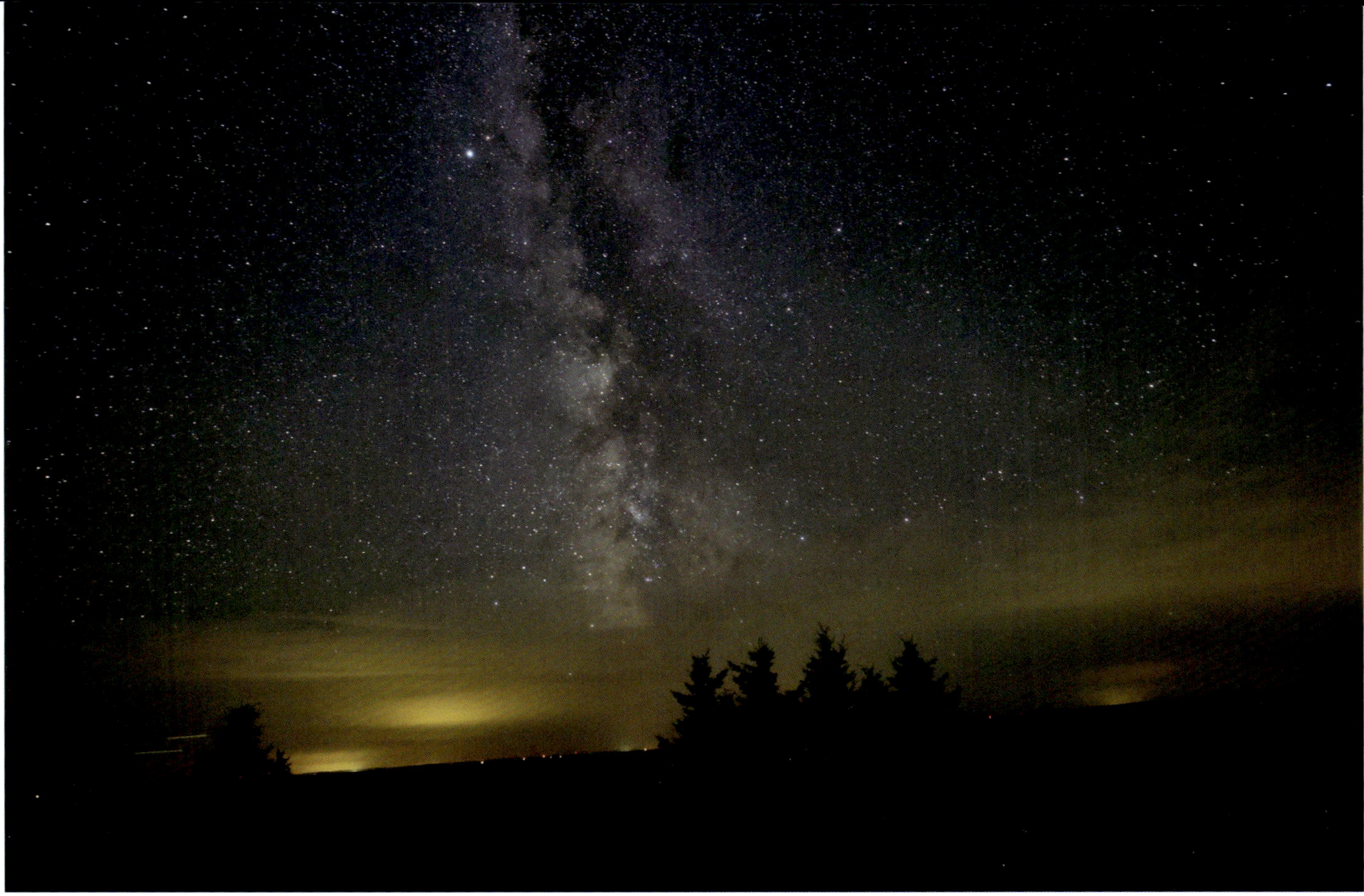

Our lights would not be restored for eleven days, and our connection to the internet for fourteen. But even through that, there was good that occurred. Our neighbours provided us with hot food and drink and checked in often to see how we were coping. My brother and sister-in-law soon came to visit and cleared away the brush from the trees bordering our lot. One neighbour gave us her cell phone for the day, and once service was reconnected, the messages of support and concern that we got warmed our hearts.

The great blessing was that there was no loss of life here in Canada, and the weather was warm. In retrospect of Fiona, we have much indeed to be thankful for.

TIM GODDARD

Where's Jimi When You Need Him?

IT'S SATURDAY NIGHT, NINE O'CLOCK. The wind and rain warnings have dropped, and it's returned to just "normal" bad weather out there. The hurricane hit thirty hours ago. We've been eighteen and a half hours without power, and I think it's going to be a while before they get things up and running again.

We knew this one was coming, and stocked up on bottled water, canned food, bread, and the Maritimer's favourite treat, *storm chips*. Some people added *hurricane wine* to the list. As post-tropical storm Fiona made her way up the coast as a Category 3 hurricane, the cone of uncertainty compressed and focused on Cape Breton. We are just west of there, and the weather reporters took great pleasure in reminding us that winds are "anti-clockwise round a low," so we would face the brunt of them.

Which we did.

A good old fashioned nor'easter, Fiona whipped in with 100+ km/h winds and gusts to 170 km/h, and over 100 mm of rain. That's a lot of rain in a short time—more than enough for a little island normally known for its potatoes, its beaches and its red-headed orphans, not for storm surges, power outages and destroyed harvests.

The previous twenty-four hours flash back through my mind: Friday night, 2:15 AM. We wake up to the venetian blinds on our windows blowing inward like sails. The latches on the windows have popped, and I climb onto a small footstool so I can access the top sills. I force the windows down, then relatch the clasps. The blinds subside, foresails in the doldrums. The streetlights show water racing in rivulets down the street below, ornamental trees whipping back and forth, a parked car shaking itself awake with flashing lights and a strident alarm. The rain is horizontal, the wind a banshee.

Suddenly there is a loud explosion, not thunder, and a purple flash bounces off the roiling clouds. A transformer has blown, the lights click out and there is nothing to see, inside or out, just the retina-memory of a purple haze. Where's Jimi when you need him? I go back to bed, lulled to sleep by the gale.

Saturday morning, 8:00 AM. A wind-whipped street, the rain still lashing the windows. I wake to a tree in the road outside, its shattered limbs splayed across the sidewalk. People emerge from their apartments and congregate in the corridor. One neighbour lost a window; it blew in at two-thirty this morning, but she got it back in place. Another was not so fortunate; the glass shattered in his living room and the floor became a sodden and dangerous mess until there was light enough to see.

We call our daughter out in the country. A tree has fallen onto her house and appears to be resting on the roofline. An apple tree has been uprooted, as have the poplar and hedgerow trees in the community pasture across the road. The dogs are getting squirrely from not being allowed outside for a proper run. She's had no power since nine o'clock the night before and is boiling a kettle on the wood stove. "I'm fine," she says, just before the connection is lost.

The phone zings with another alert from the Emergency Measures people, beseeching us to stay indoors and not to put ourselves—or our first responders—at risk. Most comply, but there are always those who believe that they are special, that the rules don't apply to them, and they walk their dogs and stroll the neighbourhood looking for images to capture. The posts on social media begin. Through the gaps in the seams of the windowsills, the wind cries Mary.

It's Saturday evening, at 7:00 PM. The phone service has stabilized now, and we're getting messages and texts from across the country. People are wondering how we're doing. Apparently, the news film from the storm has been horrendous, and people are very concerned. We have no television, and data downloading is unreliable on the cell phones, so we don't really know what's happening outside our own little isolated bubble. We light the propane burner, a relic from our camping days, and boil pasta for dinner. We open a bottle of wine. It's dark, but the storm lanterns remind us of campgrounds long ago and far away.

Sunday morning, 11:00 AM. We decide to empty the fridge and freezer and take everything out to our daughter's place. She has a generator. The main roads are clear, and there is a half-kilometre-long lineup at the only open gas station. The highway is lined with broken power poles, the wires scything through stands of corn, soybeans, cabbages. Trees are down in yards, along hedgerows, in the woods. A camping trailer lies on its back, wheels upturned, looking like a fat puppy waiting for a belly rub.

Monday morning, 8:00 AM. Friends from Ottawa email that they've just got back from Europe and seen the footage of Fiona. "Must have been terrifying to live through," they say. "What can we do to help?" I reply immediately: "The ATMs are down, and the liquor store is closed. Please send money and scotch." There is no response.

Monday afternoon, five o'clock. The island is a landscape of shattered dreams. Houses are decapitated or washed away; twelve-metre waves have produced historically high storm surges; thousands of trees are down, and the whole Island is still without power. There are photographs of fishing boats washed up onto bridges, of barns and silos exploded, of roads ending in jagged sinkholes.

The news comes in: the lowest air pressure ever recorded in Canada, the highest wind gusts for the longest time. All future storms will now be measured against the fury of Fiona. We know that these benchmarks will be challenged, perhaps not this year but in the years to come. As Bob wrote and Jimi sang, *"And the wind began to howl, hey; All along the watchtower."*

Can't wait.

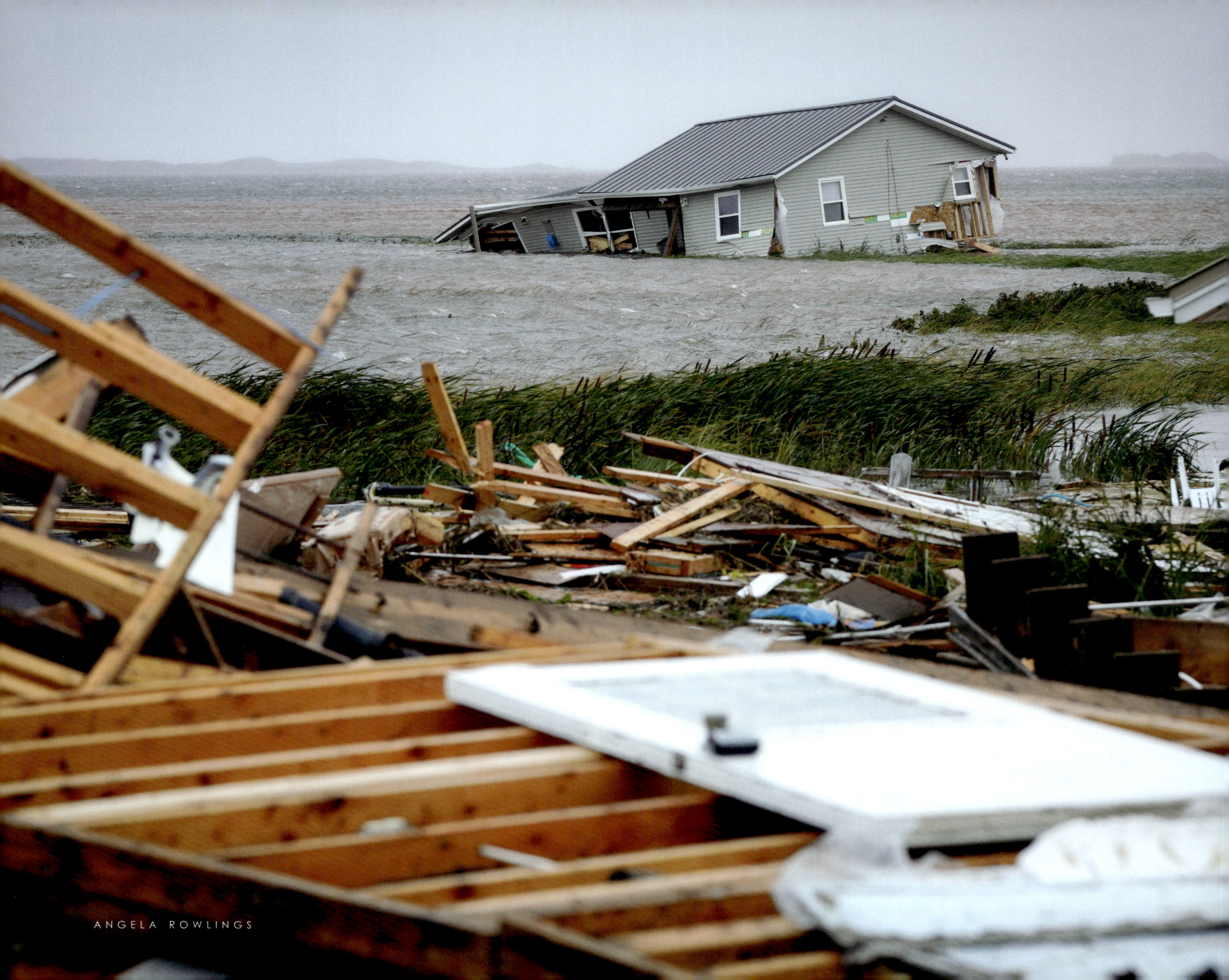

ANGELA ROWLINGS

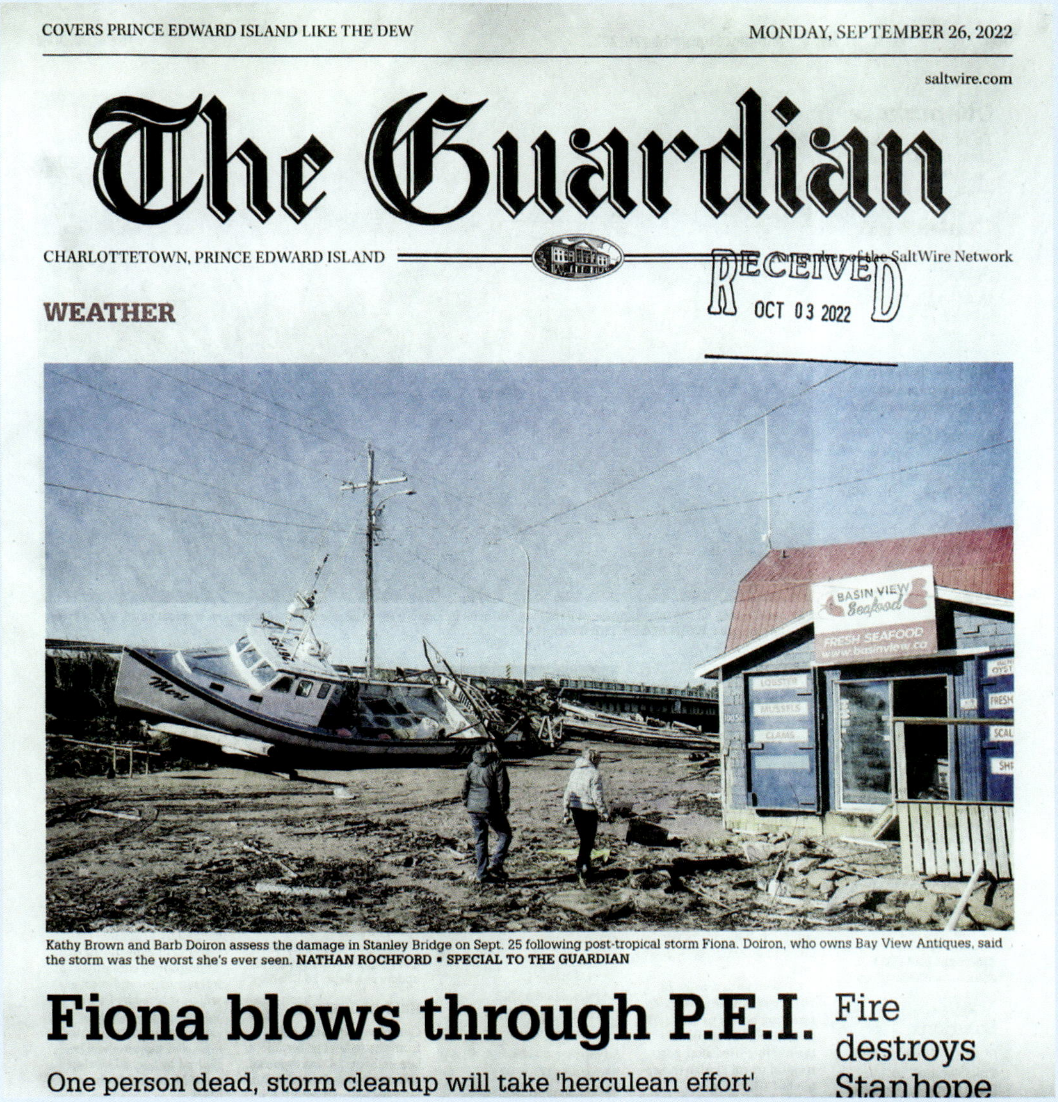

COVERS PRINCE EDWARD ISLAND LIKE THE DEW

MONDAY, SEPTEMBER 26, 2022

saltwire.com

The Guardian

CHARLOTTETOWN, PRINCE EDWARD ISLAND

A member of the SaltWire Network

RECEIVED
OCT 03 2022

WEATHER

Kathy Brown and Barb Doiron assess the damage in Stanley Bridge on Sept. 25 following post-tropical storm Fiona. Doiron, who owns Bay View Antiques, said the storm was the worst she's ever seen. **NATHAN ROCHFORD • SPECIAL TO THE GUARDIAN**

Fiona blows through P.E.I.

One person dead, storm cleanup will take 'herculean effort'

Fire destroys Stanhope

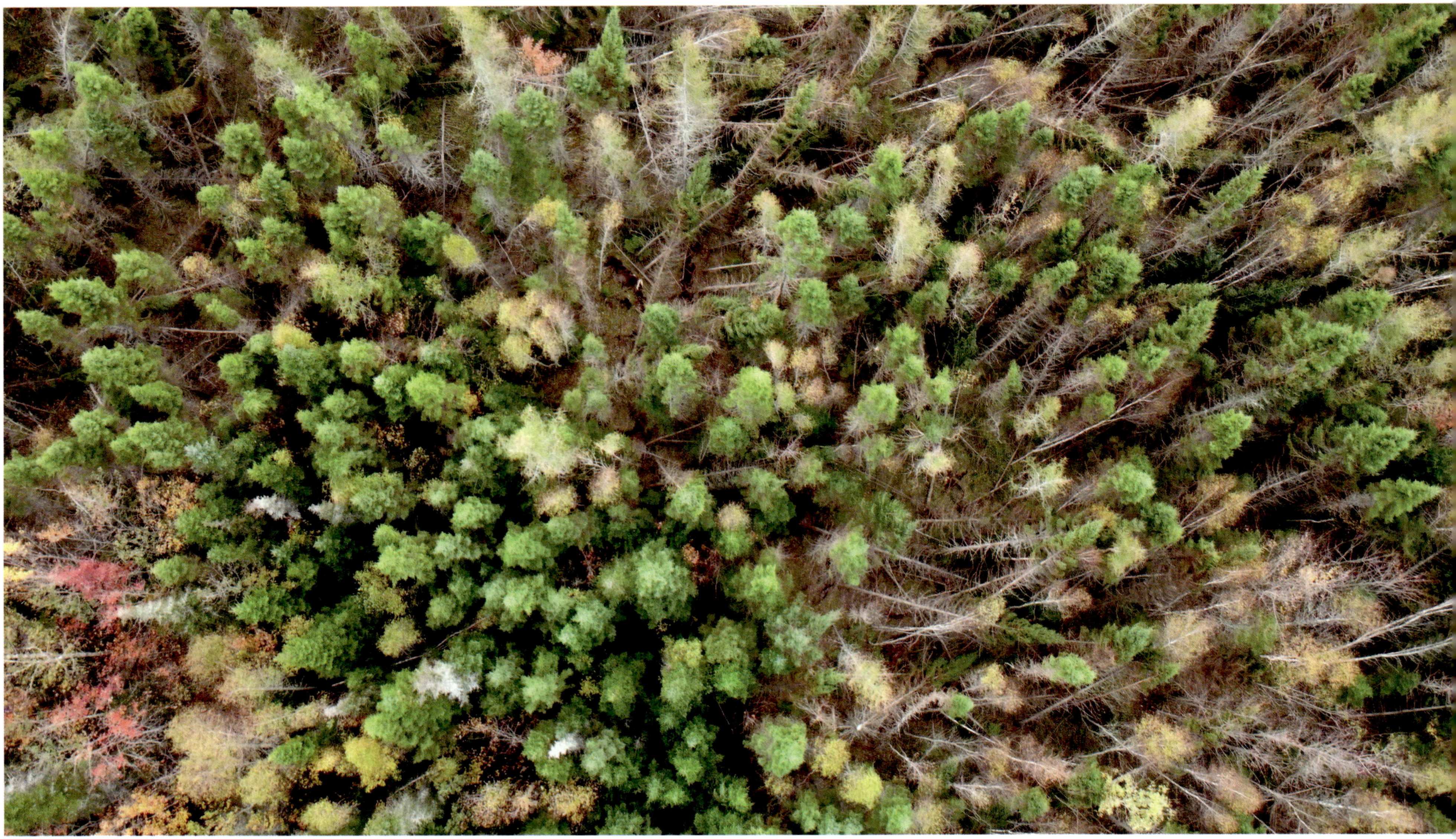

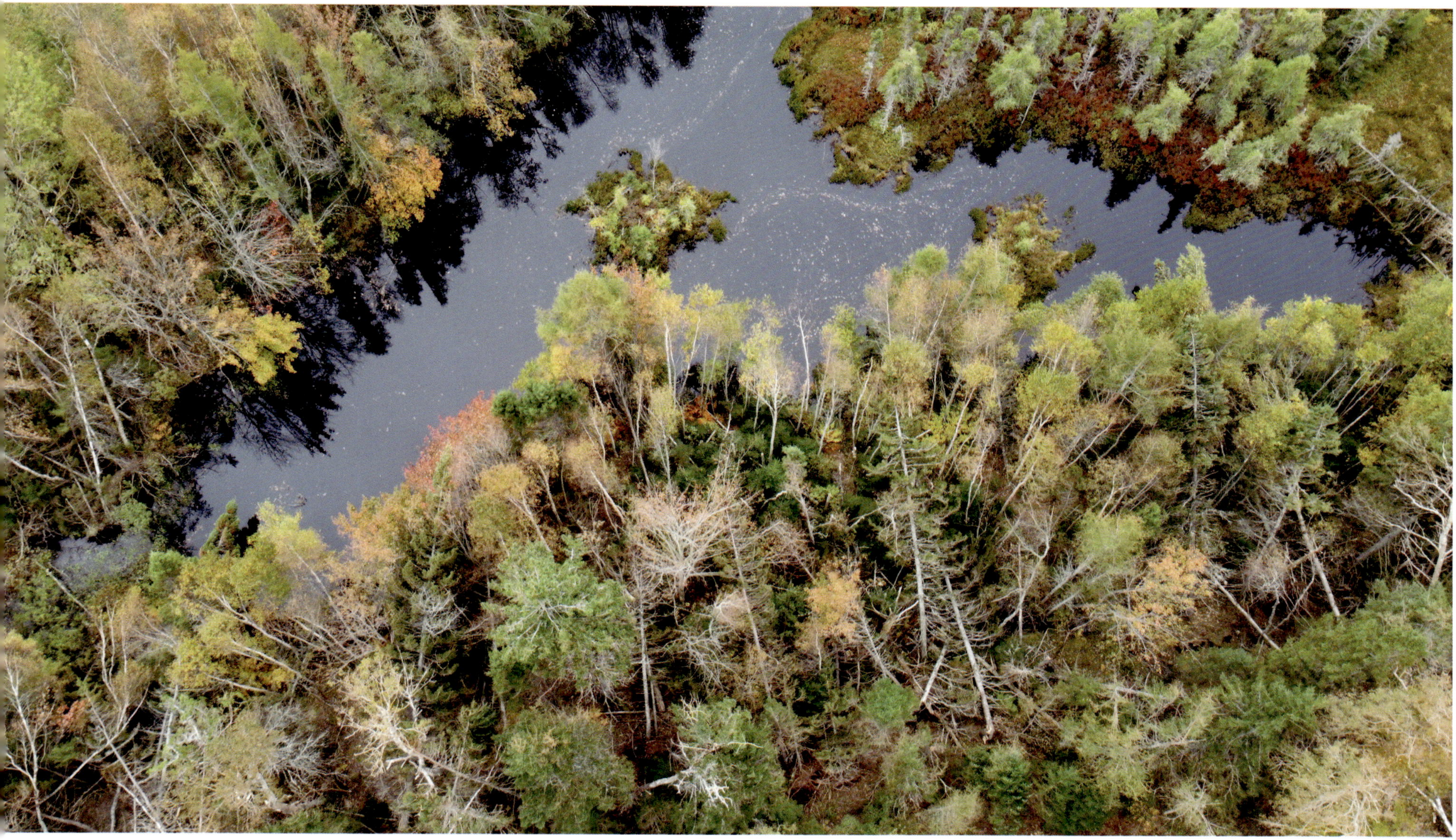

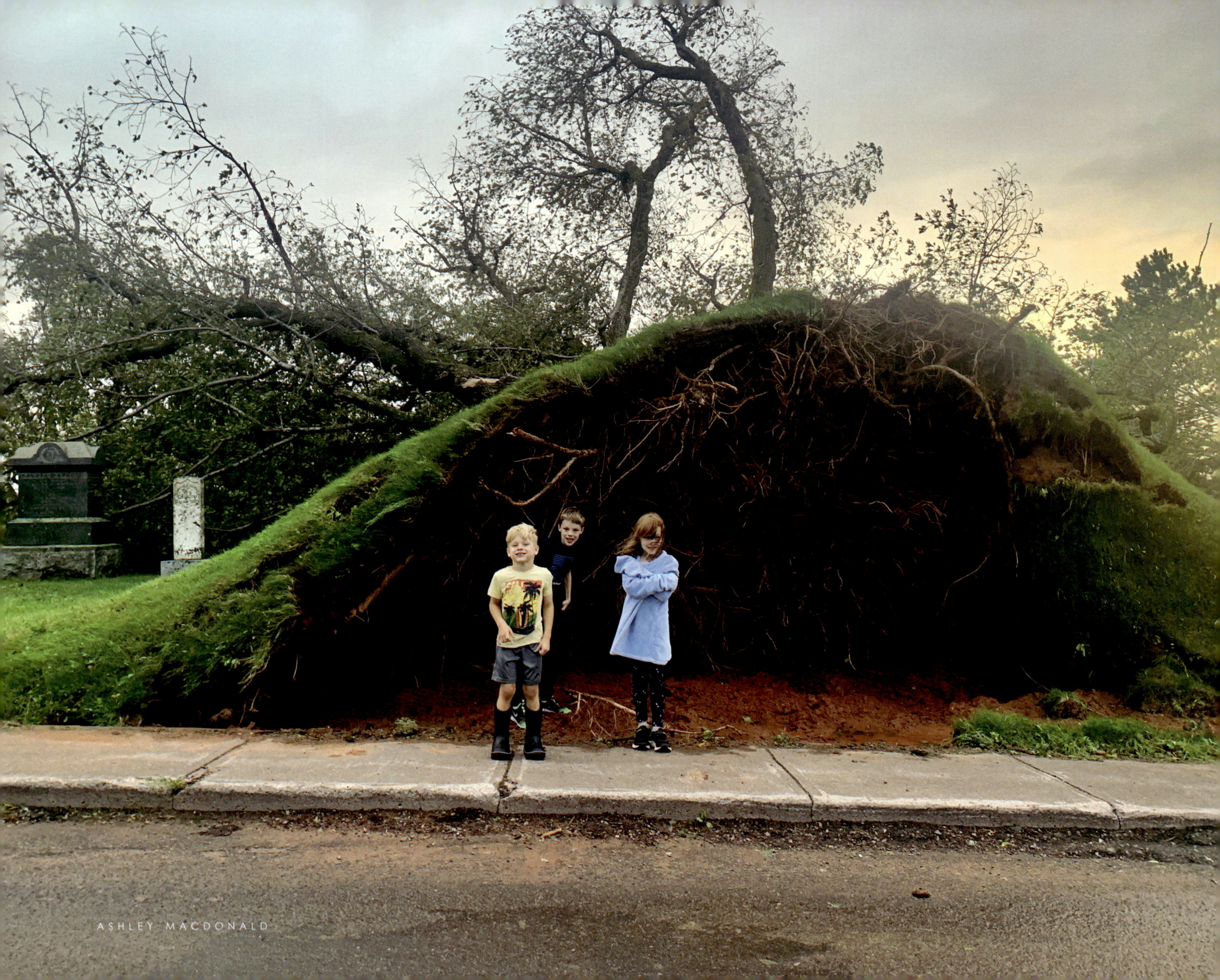
ASHLEY MACDONALD

J. M. PLATTS-FANNING

Hurricane Embrace

Fiona is caught in the eye of my hurricane
the ancient light of my soul, she gusts
green-leafed frontline soldiers make their final stand against
tempestuous wind
relinquish themselves to the root-lifting power of her torment

I feel them splinter
flecks of stoic-wood glitter on my skin, pierce maple-candied body to draw
forth sweet syrup; into deep-blue iris, a thousand thrusted glittering slivers,
sharpened, to tear down the dam; salt water gushes along meadowhawk-
kissed cheekbone, soft bee-brushed honeyed lips, as Fiona tore down
fortress-strong *oak, pine, elm, birch, ash, cedar, poplar, willow, maple...*
fertile-apple ~

~ apples filled the sky
suspended sweet fruit debris
plucked from the vortex air
their taste could never be as sweet *as gifted mantra messages crossing radio*
frequencies

Fiona's special forces sought out the secret cave-eye of my aching heart
blew inescapable wind inside; swirled the conflict
broke down the thin, jagged tissue wall so it couldn't be ignored any longer
cleaned out my dusty heart
and all that surged outside
revealed what the tarot witch knew to be true
and I wept, harder than Fiona could bleed water

ribbons of wind
created by the stirring of his hand, on her crown; fingers lightly swim
through waterfall hair
sliding down between soft red sheets
down curved dune back
under sheltered skin
just a simple request
a scented embrace to last a lifetime

trees lean south, on addictive electricity
a message
turn off, tune in, look up

Orion spilt the Milk he carried across the cosmos to light the Way for the ones in darkness
headlamp light cast through blue water bottle made diamonds shine on the ceiling;
and in the darkness
shadowplay filled the walls
at first, maternal
but then ~
something else crept
something flickered
in the dense bleak of cold insomnic-night; erotic images licked the walls

the birds raged through pink sky
grieving for their trees
sharing their hurricane stories
swerving passed crumpled transformers, homeless waterlogged mattress,
split poles, knocked-over electric dominos
bees and wasps awoken after the storm by rays of out-of-season heat
disturbed by chainsaw symphonies and the sunrise-timed orchestra of rumbling generators
tree zipper slashed opened red-clay island.
sandstone teacups smashed against cliffs spill sacred sea foam memory particles

one rarely gets to touch the tips of royal tree crowns
but now I lay within brilliant arms that spent a hundred years reaching for the sun
now they reach for me
a final embrace to last through the slow dying days ahead
walking through horizontal spruce curtains
the bee brushed against my lips and I felt the world spin me an inch
re-orienting my gaze to a green frog clutched to upturned Celtic knotted roots, revealing a teaching
an as yet, indecipherable symbol, clutched ~
on the edge of human perception

desperate for Earth's embrace
curled in a patch of sun-warmed moss
I am the woods decomposing
allowing for new growth
healing from the hurricane

here, I am not human
but something *belonging*
a long, two-legged Ent[1]
with twigs knotted tightly to create emerald diadem, *its crowning jewel*
~ a camouflaged gray inchworm

---

[1] A species of beings in J. R. R. Tolkien's fantasy world, Middle-earth, who closely resemble trees.

I am Fiona howling
for the grave marker trees
for the City Giants
the Old Growth Behemoths, which stood watch, as a city was born and shaped around them
I wept for my own longing
to be rooted in this wood
to feel what the tree feels underground
under
Mother
inside her womb of expanded consciousness

I've seen hurricane loneliness fill the cracks of time-weathered skin
felt the yearning tears roll down my neck of a stranger left shoved in a dark corner with only the company of smooth wooden Moai heads and death gods turned to face the walls of elderly hell for seven days, reflections of afternoon light bouncing off youth trapped in dusty frames, proud soldier looked down, and asked, *what was it for?*

I've seen hurricane battle-wounds reveal neglect

gauze-bathed a drunken blue-eyed angel-aged, whose flirtatious blood was dried like the cement he'd smashed whiskey head against, after days of being left to wither and harden in food-spoiled government-funded housing squalor, only thing for pain and empty stomach is alcohol-fuel, would exchange flashlights and cooked chicken for a cold shower with a warm-blooded woman.

sometimes, resuscitating kindness is pouring a drink for a wrinkled forebear, with the shakes, whose weathered many-a-salt, mind-flaying-storms;
sometimes, resuscitating kindness is a five-dollar strawberry milkshake, and ~

~ *a warm hurricane-engulfing embrace.*

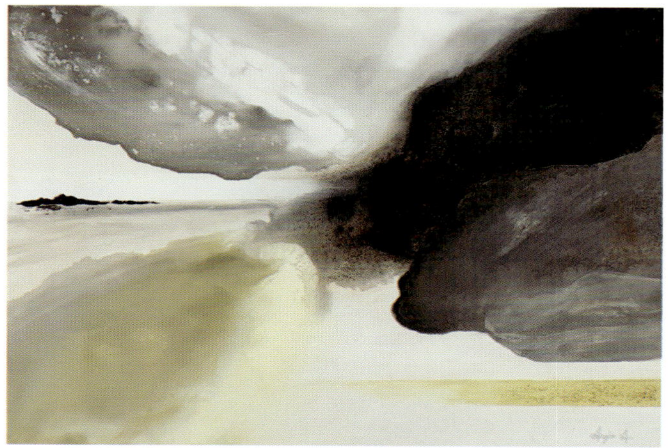

ANGIE ARSENAULT

SEPT. 24, 7:00 AM — Looking out the window, it takes a minute to fully comprehend the view.

SEPT. 25 — We are well, and feeling very lucky. Trees are down all over the property, snapped, tipped, roots the size of sheds. I lighten my mood by thinking I'll make a few art installations with the cleanup. The forest looks like a museum dedicated to the power of wind.

We are grateful for a generator and the use of limited items: hot plate, a freezer, a kettle. It's funny how quickly you know what the essentials are.

We spend the morning dealing with debris. Amidst the devastation, I notice the violets that live near the back steps, called heartsease. I have no idea how such things survive. Emotions are raw, and small things are precious. And then I see Shelley, our resident monarch (hatched under the outside tap), who I can't believe stuck around for this. A crow flies overhead in its usual pattern, lower here, circling there, preparing to land on its favourite branch, but the whole tree, an eighty-foot spruce, is gone, on top of the carport now along with a few others. The crow isn't amused, but it will adjust. We all will.

I walk around with a flashlight, and a neighbour comes by to make sure all is well. I'm touched by this small exchange. We connect with other neighbours too, some of us meeting for the first time. We talk of who needs what, and how each one can help. All I can think is how extraordinary people are.

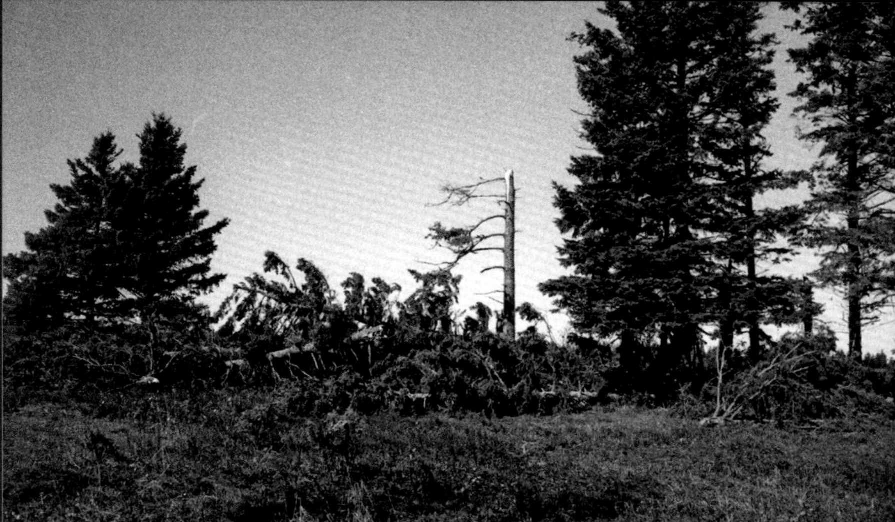

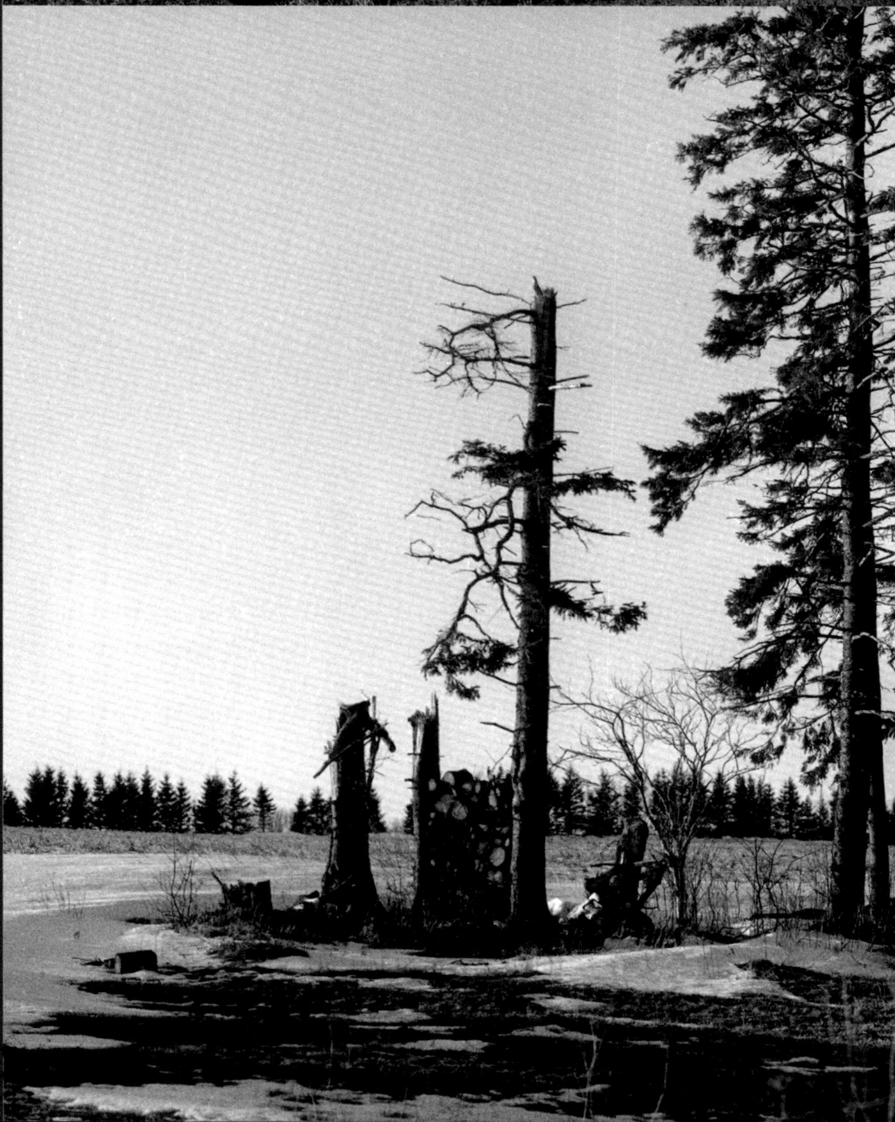

LISA CHANDLER

No Ground Beneath My Feet

When things fall apart and we're on the verge of
we know not what, the test of each of us is to stay
on that brink and not concretize.
The spiritual journey is not about heaven
and finally getting to a place that's really swell.

—Pema Chödrön
When Things Fall Apart: Heart Advice for Difficult Times

IN THE WEE HOURS OF SEPTEMBER 24, Hurricane Fiona, figuratively and literally, took the ground out from under us. Parts of our Island fell apart. Parts of me fell apart too: my outer and inner landscapes have been forever altered. *Until Fiona, I'd felt near death twice.*

The first time was years ago, in Northern Mexico, when I stupidly went scuba diving without training. I got into trouble that I didn't know how to navigate. Though my panic only lasted a few minutes, the memory of feeling like I was drowning is forever etched on my brain.

The second time was during childbirth. My doula Francine had warned me that I might feel like I was dying during the "ring of fire" stage of labour, but assured me this was a wonderful thing, as it would mean that my baby's head was crowning and that I'd give birth soon after. True to her word, I did feel near death at a certain point, after a long, induced labour. I remember freaking out, and then digging deep, to push through. I credit her sage advice and birth-coaching for getting me through that painful and frightening stage. Shortly after, I was gifted with a beautiful baby girl.

Enduring Fiona was different.

My Fiona fear has had a long half-life. Unlike my earlier, brief "near-death" moments, my fear over Fiona started in the days leading up to her arrival, and lasted for many punishing hours during. In the weeks after, when I was able to find sleep, I had unsettling dreams. The rest of the time, I'd lie awake worrying about future storms.

Other than the routine preparations, I knew I had no control. We did what we could; then, I had to surrender. I spent the night believing we might die or be seriously injured. The contents of my stomach turned to liquid, but going to the bathroom seemed like a death-defying act. Each time I braved

ART BY TOMOYO SUZUKI
COURTESY OF THE AUTHOR

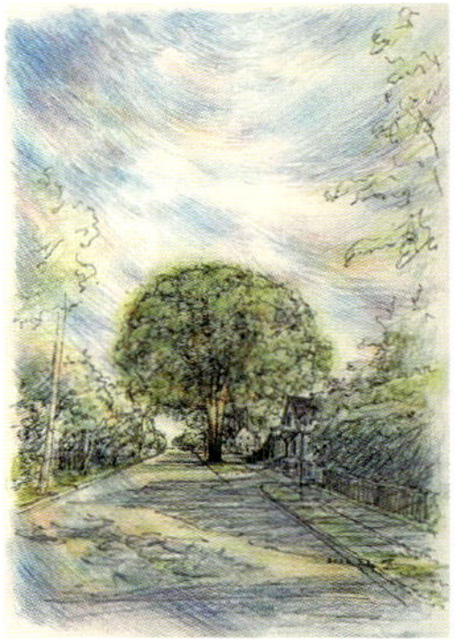

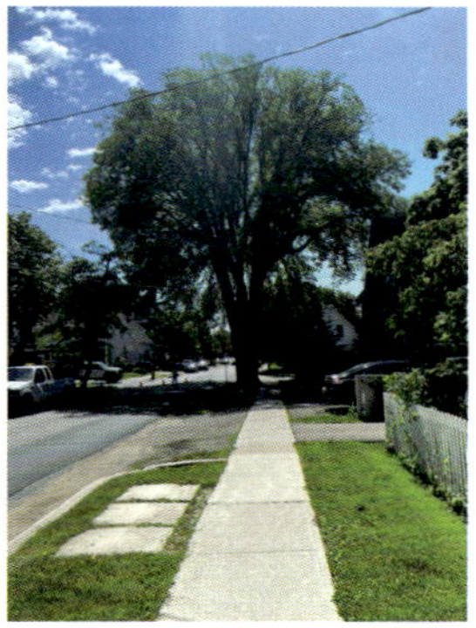

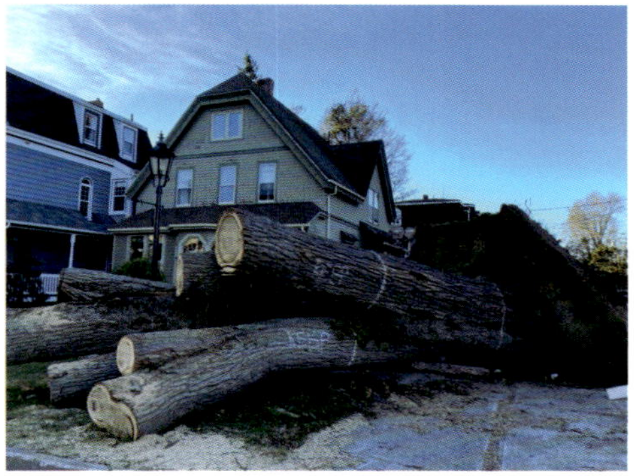

a trip, I'd watch in fear as our neighbour's tree bent practically in two with every gust. And while it didn't fall on the house as I feared, it did break across the roof of our detached garage. I felt fortunate not to have seen it happen.

I felt very alone, and I got through the night by texting my partner, Peter. With autism and contagious emotions in our kids, we knew we had to avoid potential storms inside the house by being separate. It was hard. In the days leading up to Fiona, I was angry about it. The night of the storm, I got through by breathing and hugging my (thankfully) sleeping daughter tight. Each time I got totally wrapped up in my own fear, I did my best to zoom out to all the people I love and wish the best for in a Metta-like meditation; I'd learned it during fertility treatment days when an excellent therapist advised me that when I felt resentful of pregnant moms, I could feel my sadness and jealousy and choose to shift from my self-focus to wish the best for others. It helped!

It was all punishment and no reward. There was no beautiful baby placed in my arms at the end. Of course, being alive and uninjured was not a small thing. That there was no loss of life from Fiona's impact seems a miracle still. There was, and still is, devastation, and I see more every time I venture to new areas.

And while I don't know what each endured during and after the hurricane, I imagine everyone has harrowing stories to tell. And I also imagine many with feelings pushed down. Might you, like me, be grasping for how to control things to avoid this suffering again?

As I opened, I will close: Pema reminds us that life is a good teacher and a good friend. Things are always in transition, if we could only realize it. Staying with the shakiness—the broken heart, the rumbling stomach, the feelings of hopelessness—is the path to waking up. Learning to relax in chaos, not to panic, and to gently and compassionately catch ourselves when things get edgy; it seems like that might be Fiona's message to me. Let us practice peace with ourselves and others.

MÉLANIE MORIN

Blown Away

FIONA HAS BEEN RAGING ALL NIGHT, and sleep is evasive. The power went out at 1:00 AM, and the loud cracking noises in the dark of night are frequent and ominous. The wind is still trying to infiltrate the airtight windows on my house, and a soft whistle sound tells me she is having some success. The house shakes occasionally, and with a loud crack, I am aware that something has hit the house—but the force of the wind prevents any possibility of opening any doors to have a look. But as the sun rises, Fiona's surprising acts of devastation are illuminated, and I begin to see.

The windows are covered with a film of salty water and tree leaves that are chewed up into tiny pieces, and which polka dot every surface. The leaf bits have mixed with rainwater on the street, creating a slippery green sludge. Where my backyard used to be, I find piles of trees, branches, random shingles, and other items that flew in from elsewhere. My neighbours' brand-new fence is now propped up against the side of my house, completely flattening the downspout.

We wait things out as the gusts are still blowing, but the dangerous winds begin following their tracked path away from us and on to Newfoundland. At around 2:00 PM, I start to see people come out of their houses, and Jamie and I decide that it might be safe enough to venture out to the street to get a better sense of what Fiona has left behind in her wake.

We see some trees down on the street farther up, but as we come around the corner onto Greenfield Avenue, the scene makes us stop in our tracks. A giant tree is strewn across the road, roots poking out of the ground as high as two people, taking down the pole and power line with it. Across the street, I see another set of giant roots and the massive tree they belong to, fallen across two properties.

We stand on the sidewalk, assessing our next move. I see a crow on the ground nearby. Crows, who often gather by the thousands in our neighbourhood trees in the evening, are loud and arrogant, but this crow staggers, unrecognizable. He is bleary-eyed, dishevelled and stunned, deeply humbled by the power of the hurricane.

We see a few other people coming around the tree on the road, gingerly stepping over the power lines on the ground, so we head in that direction. As we get closer, we see two cars under the branches, their owners surveying the damage. Once we get past this first tree, we get a better sense of the extent of the destruction. The entire street is strewn with dozens of trees, their whole root system exposed vertically. Farther up, I notice a cat walking across a horizontal trunk, perhaps curious about how gravity had seemingly shifted overnight.

The more we walk, the more sadness seeps in. We can see the grief in the eyes of everyone we pass as they look at the mature trees uprooted, shaking their heads with disbelief. The loss of the many trees that made up the canopy of our streets hurts deeply and unexpectedly.

We get to the end of Greenfield Avenue, assessing the state of each property. Most have lost several trees, and almost all the houses have sustained damage from the winds. As we reach Brighton Road, we see that the path into Victoria Park is completely blocked off by fallen trees, and some of my favourite century-old trees are broken apart and lying sideways across the road, taking out the beautiful streetlights and pulling up the sidewalk. Dangling wires make getting around these trees seem dangerous, but we continue, committed to being first-hand witnesses to the devastation.

We pause and take pictures along the way, each broken tree more devastating than the next, until we finally reach the boardwalk. While the park is a cluttered mass of fallen trees to the left, we feel a sense of relief that the boardwalk is mostly intact.

Finally, we come around and re-enter the neighbourhood on Ambrose Avenue. Near the first corner, a house sits with two large trees leaning up against it, damage visible on the roof. One of the trees has uprooted everything around it, to a point where if you remain on the sidewalk, you are going uphill around the corner. Several power lines are dangling low or lying on the ground in the intersection, left on the ground like skipping ropes.

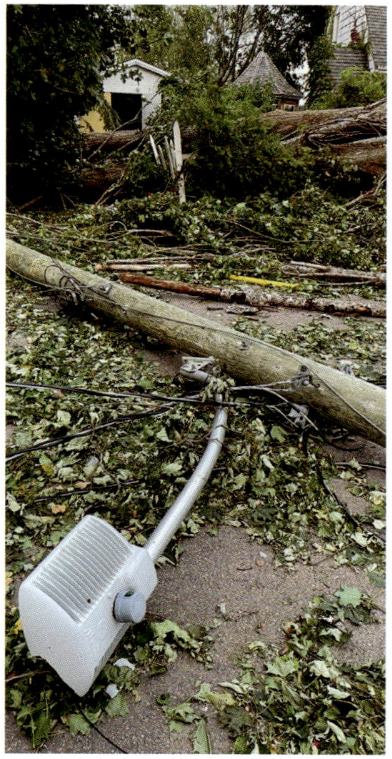

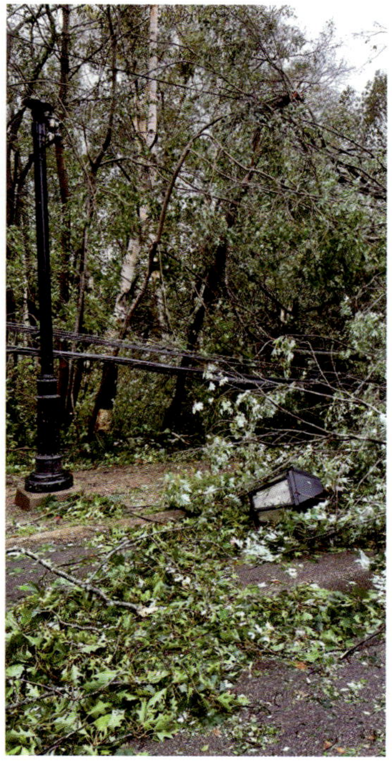

The wind is still blowing steadily, and we continue our treacherous journey. But we are stopped in our tracks once again. This next scene is by far the most surprising sight of the day. A spotless lawn—not a branch, not even a leaf, just freshly raked grass. Its pristine condition is simply stunning among the chaos and devastation that surrounds it. An older man with long white hair and a rake in his hand places small branches onto a neat pile by the driveway, erasing the very last trace of Fiona on his property. We stand there for a minute, staring dumbly at it in our confusion. What in the world? We can't help but giggle, then break out into laughter.

This one man decided to single-handedly show us how to get past Fiona. By his actions, he was saying, loudly and clearly, *Don't stand around and cry over your fate; pull up your sleeves and start cleaning up the mess already.* And so we did. We went home, and the next day—and the weeks after that, despite our grief—we got to work and cleaned up together because this giant mess of fallen trees and power lines wasn't going to clean itself.

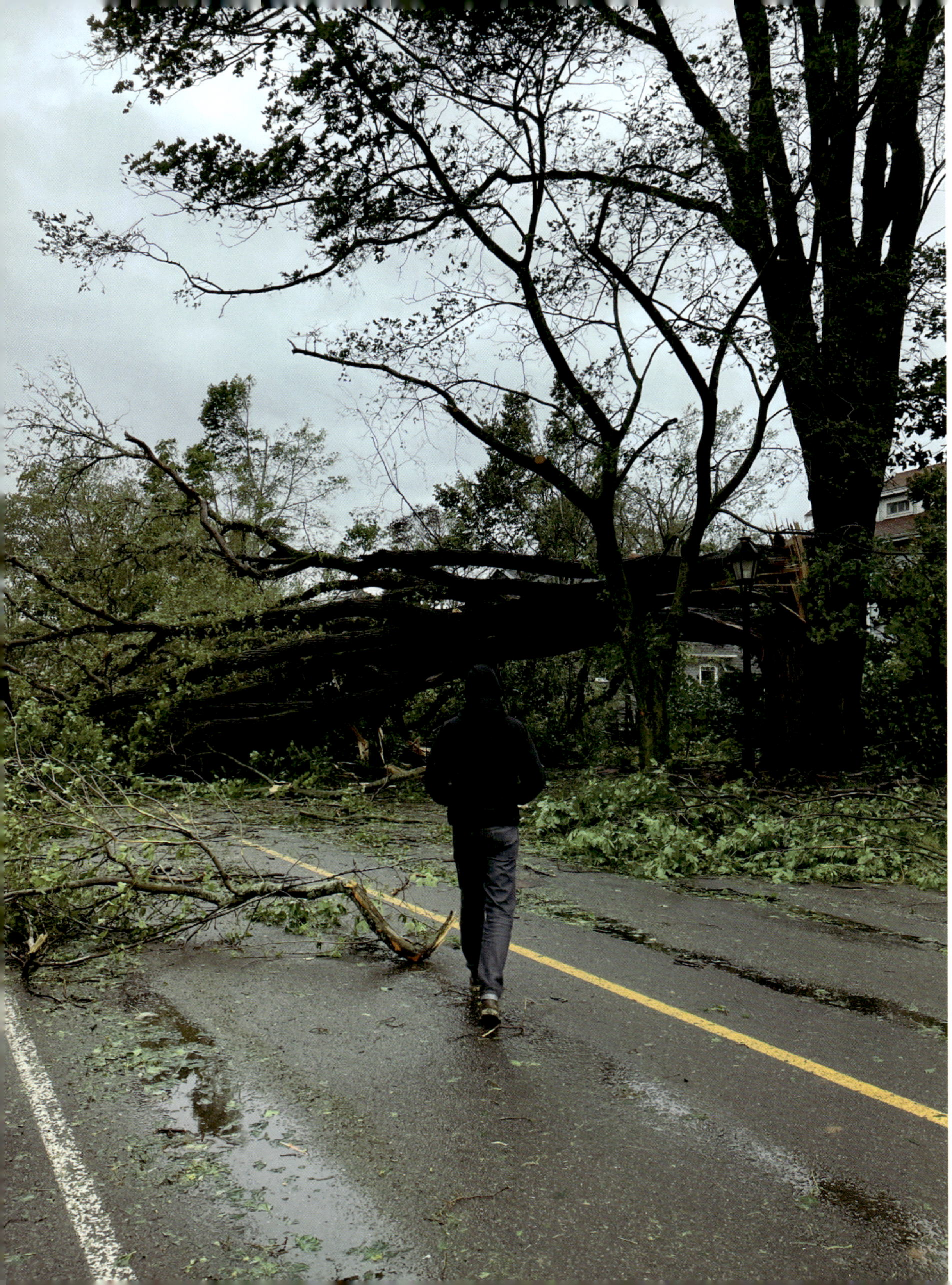

MÉLANIE MORIN

KATE DEMPSEY

The Park Bench

The winds came
Trees fell and dunes washed away
Landscapes shaped slowly through years of shifting tides and ocean breeze
Forever changed

Amidst the debris, a park bench
Flattened by its shady protector
Uprooted in the storm
In finer weather we sat on this bench, you and I
Observing goings on, telling stories of cabbages, of lives lost too soon
Exploring love, and sorrow, and safety
Each other

We sat on our bench in the dark
The night before I released my heart from its protective casing,
Exposing it to you, to us, to the elements

The landscape of my heart, shaped slowly through years of learning to trust,
To love, to be
Also changed in that storm
Broken by the weight of external forces
Like our bench collapsing under the trunk of a 100-year-old maple

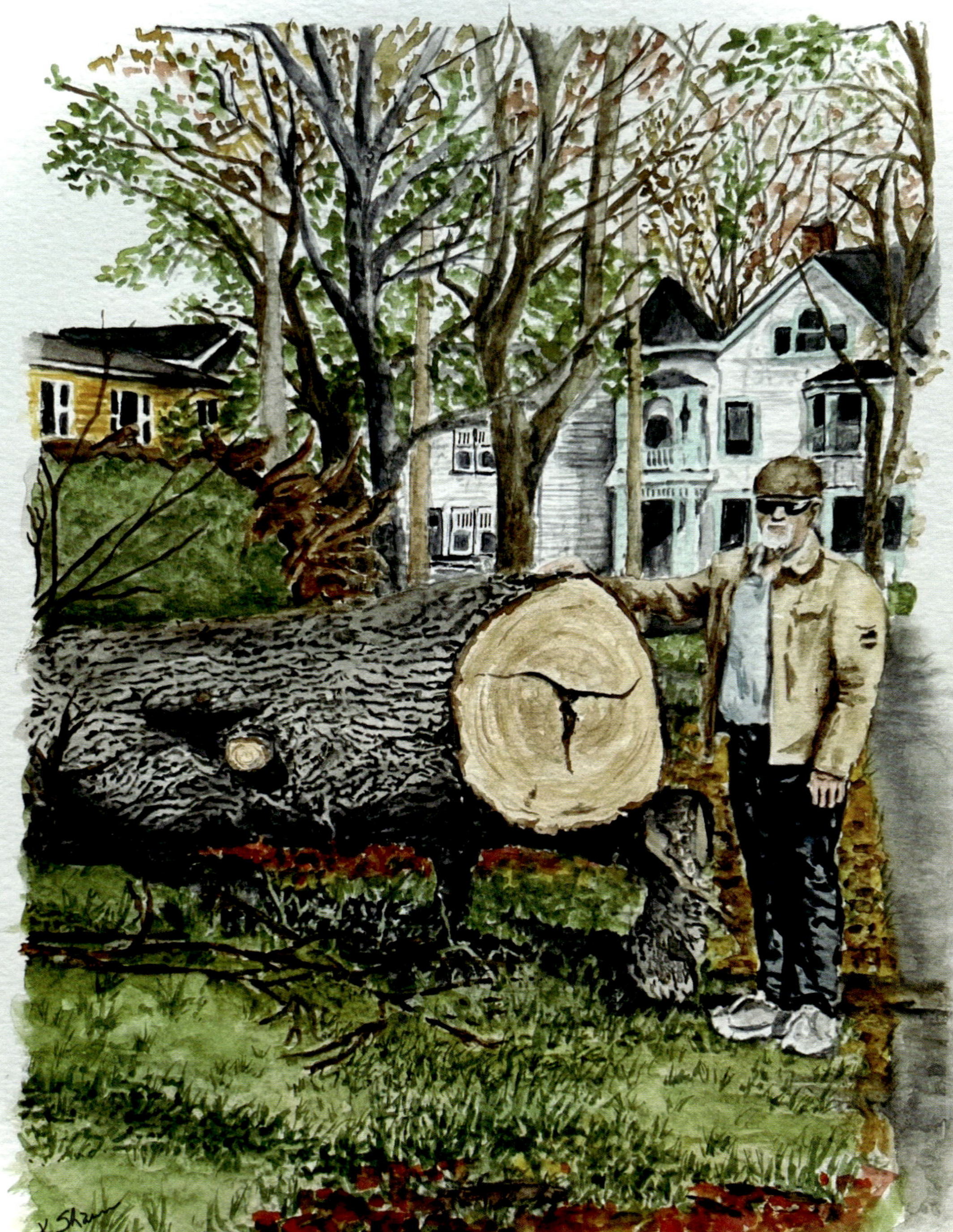

K SHAWN LARSON

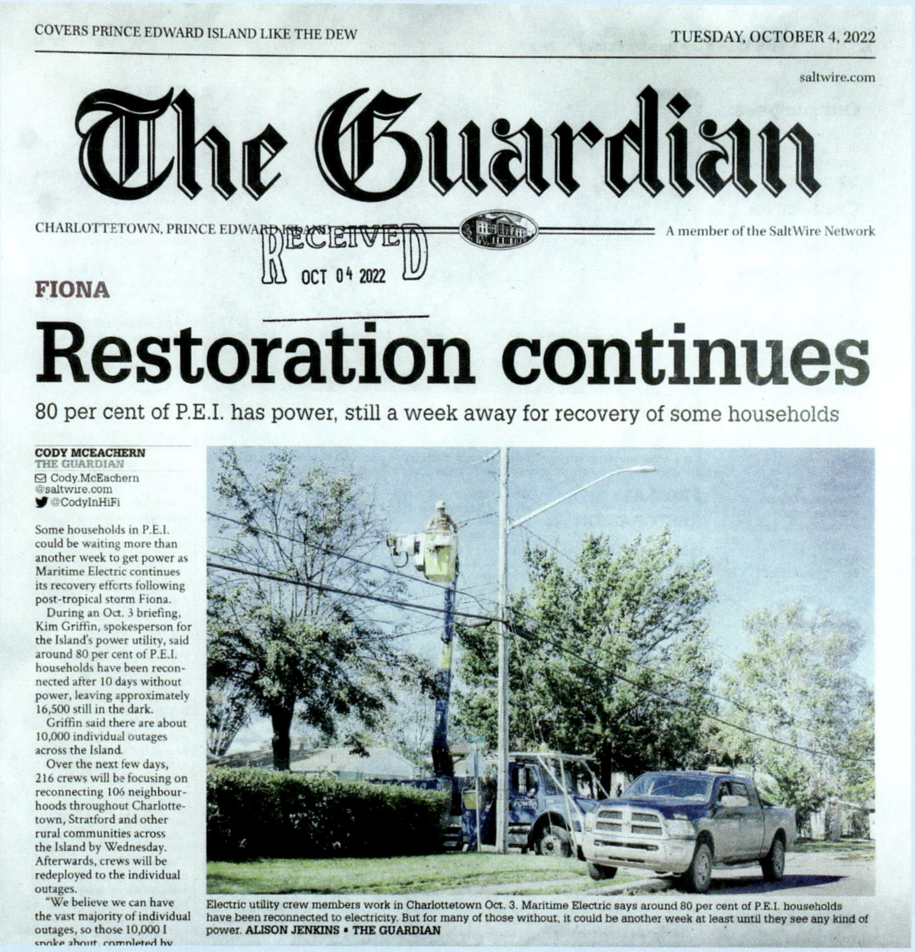

COVERS PRINCE EDWARD ISLAND LIKE THE DEW — TUESDAY, OCTOBER 4, 2022

saltwire.com

The Guardian

CHARLOTTETOWN, PRINCE EDWARD ISLAND — A member of the SaltWire Network

RECEIVED OCT 04 2022

FIONA

Restoration continues

80 per cent of P.E.I. has power, still a week away for recovery of some households

CODY MCEACHERN
THE GUARDIAN
Cody.McEachern@saltwire.com
@CodyInHiFi

Some households in P.E.I. could be waiting more than another week to get power as Maritime Electric continues its recovery efforts following post-tropical storm Fiona.

During an Oct. 3 briefing, Kim Griffin, spokesperson for the Island's power utility, said around 80 per cent of P.E.I. households have been reconnected after 10 days without power, leaving approximately 16,500 still in the dark.

Griffin said there are about 10,000 individual outages across the Island.

Over the next few days, 216 crews will be focusing on reconnecting 106 neighbourhoods throughout Charlottetown, Stratford and other rural communities across the Island by Wednesday. Afterwards, crews will be redeployed to the individual outages.

"We believe we can have the vast majority of individual outages, so those 10,000 I spoke about, completed by

Electric utility crew members work in Charlottetown Oct. 3. Maritime Electric says around 80 per cent of P.E.I. households have been reconnected to electricity. But for many of those without, it could be another week at least until they see any kind of power. **ALISON JENKINS • THE GUARDIAN**

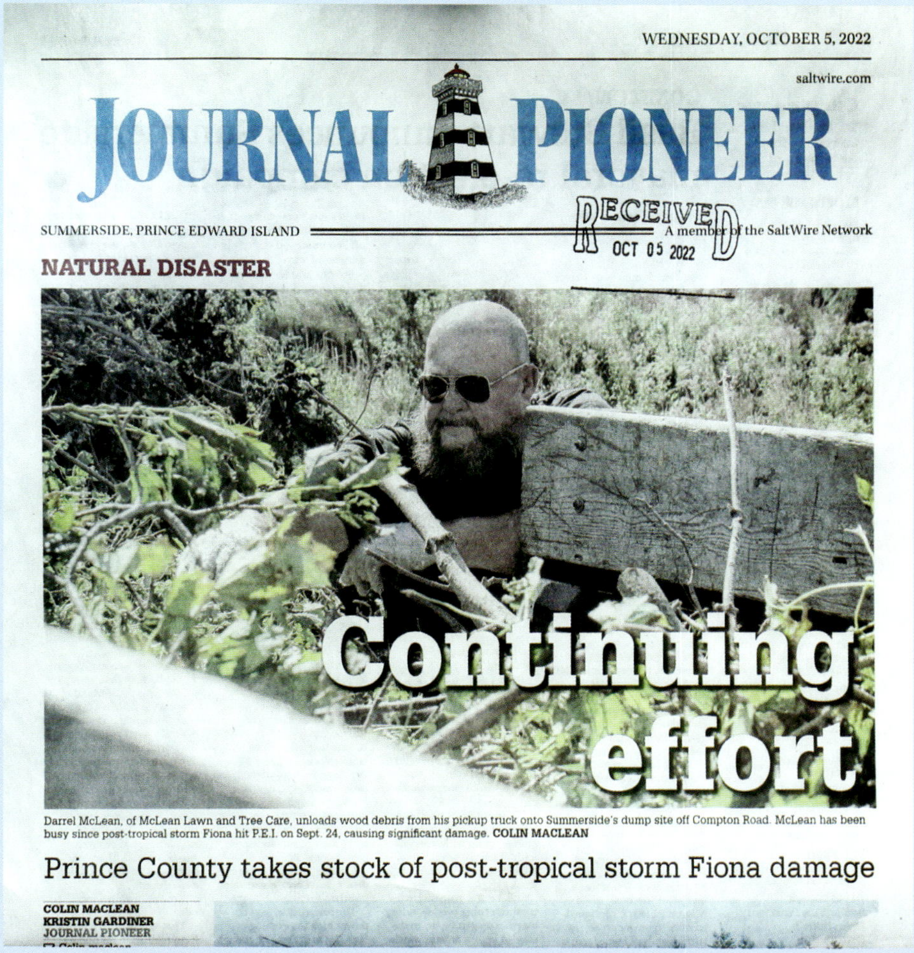

WEDNESDAY, OCTOBER 5, 2022

saltwire.com

JOURNAL PIONEER

SUMMERSIDE, PRINCE EDWARD ISLAND — A member of the SaltWire Network

RECEIVED OCT 05 2022

NATURAL DISASTER

Continuing effort

Darrel McLean, of McLean Lawn and Tree Care, unloads wood debris from his pickup truck onto Summerside's dump site off Compton Road. McLean has been busy since post-tropical storm Fiona hit P.E.I. on Sept. 24, causing significant damage. **COLIN MACLEAN**

Prince County takes stock of post-tropical storm Fiona damage

COLIN MACLEAN
KRISTIN GARDINER
JOURNAL PIONEER

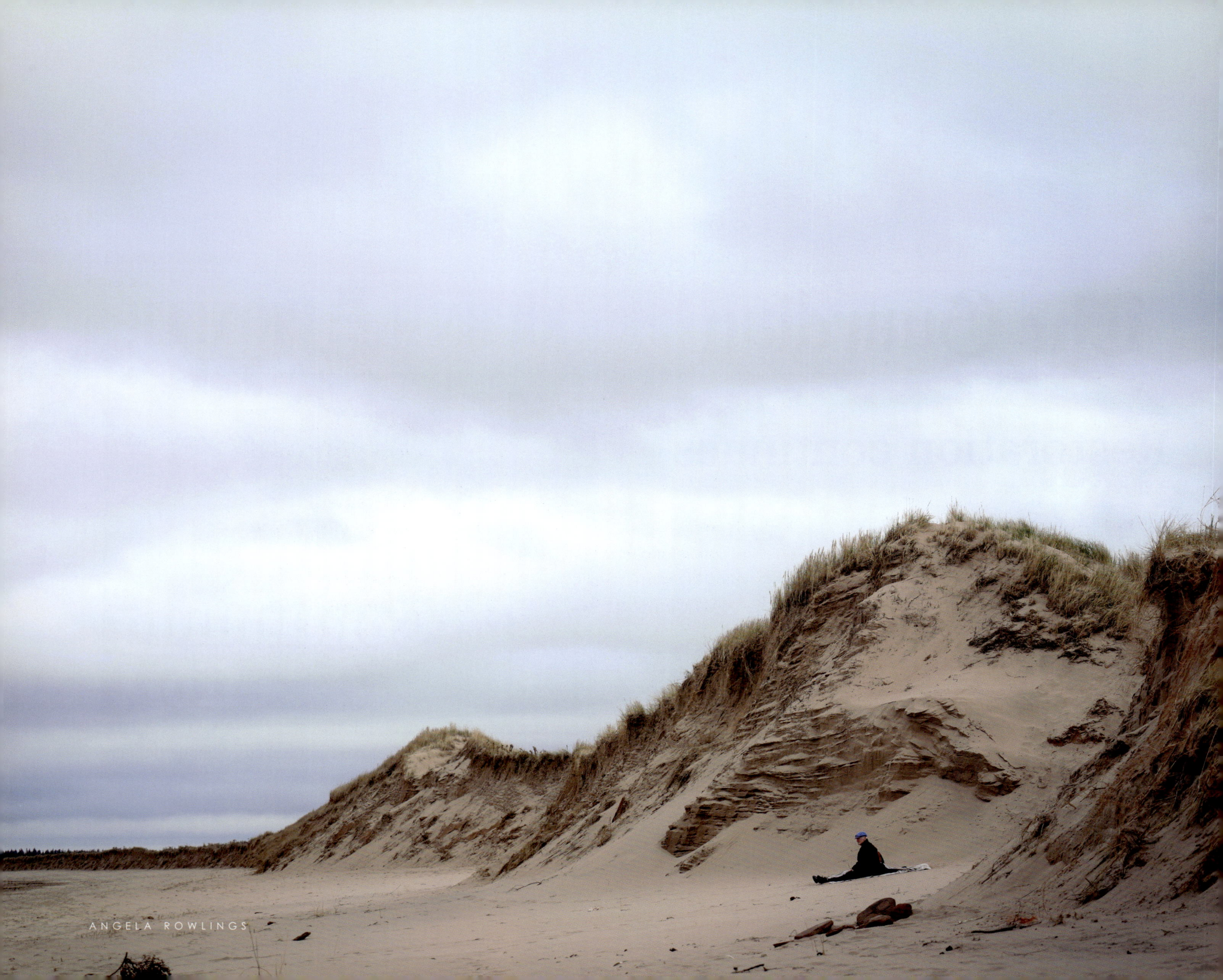
ANGELA ROWLINGS

Roots

JORDAN BEAULIEU

i.
dawn breaks on the beaten skyline
night ends sleepless, spent and small
September's worst winds carry
decades of patient presence low
roots—fresh and thin—suspended
airborne soil sweetening
days afterward

ditches laden with cable
miles of missing dune
tides tour coastal gift shops
souvenirs lively in the surge
jerrycan of heroic vermilion kisses
pumps potent to preserve
matter unstilled

city folks' steady styrofoam cups
braced low on marble floors
volunteers swiftly cover
radiant tinfoil trays
warm blankets to ward off
scarcity, perilous and lonely
dogs eager at our ankles

landlord's boots scrape the roof
army arrives and vanishes
radio returns to regular programs
machines hum a long song
trees graft, waxy and well
branches bare shedding fallen
fruits free at the orchards

ii.
weathered cedar
shaped to a seaside home
scented by weepy blossoms of rot

hallowed sweet and stale
upon an island of soft stone
shrinking, steadied
by unclad roots
weeds, aster, forget-me-nots

amongst wild bouquets
PVC passageways bloom
from earth sunbaked and crumbling
pilgrims impregnated by salt
emerging triumphant
from the cliffsides

when the tide comes
we'll rise with it
upon an island of debris
shining under the sun

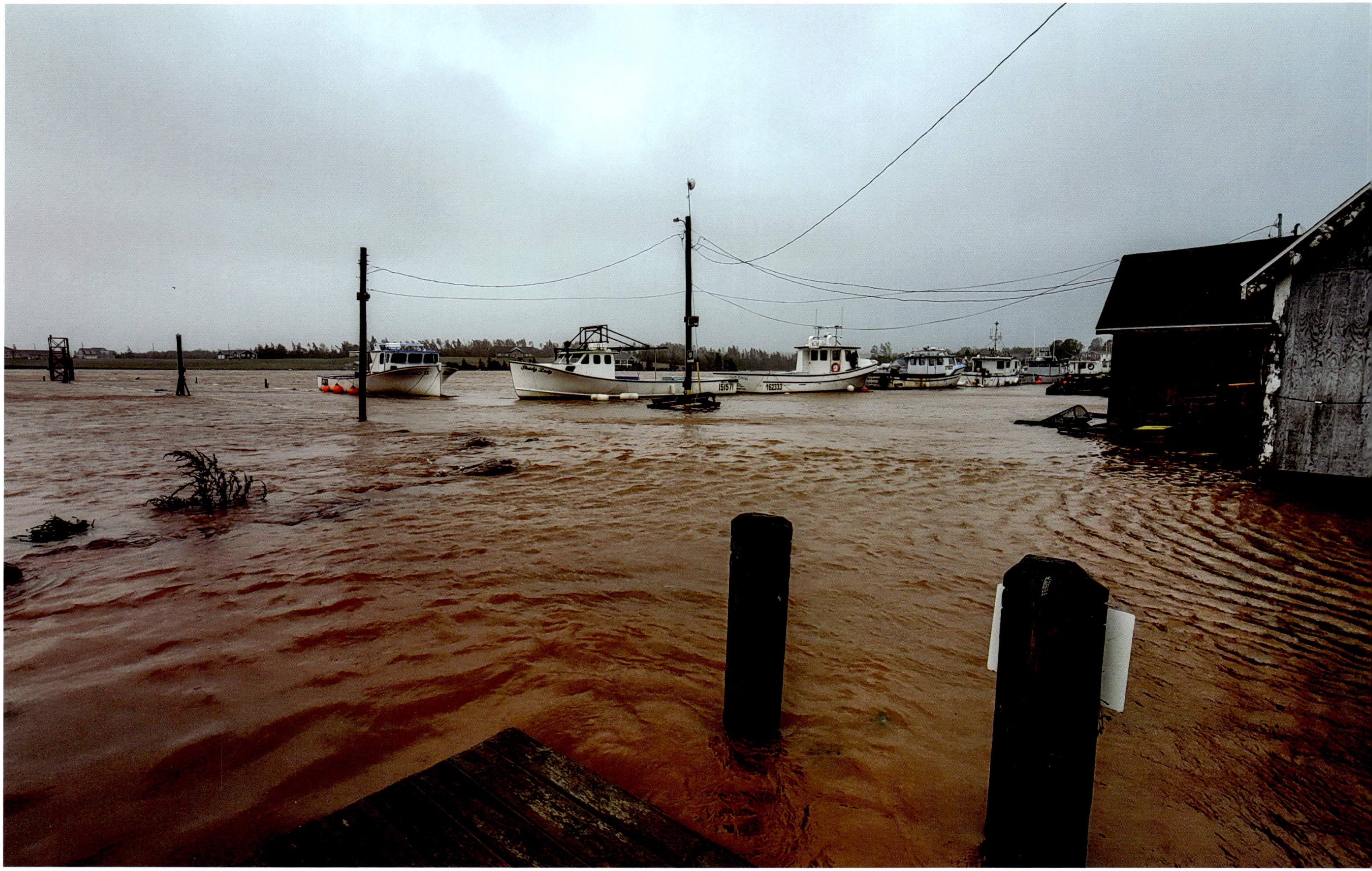

151571
162333

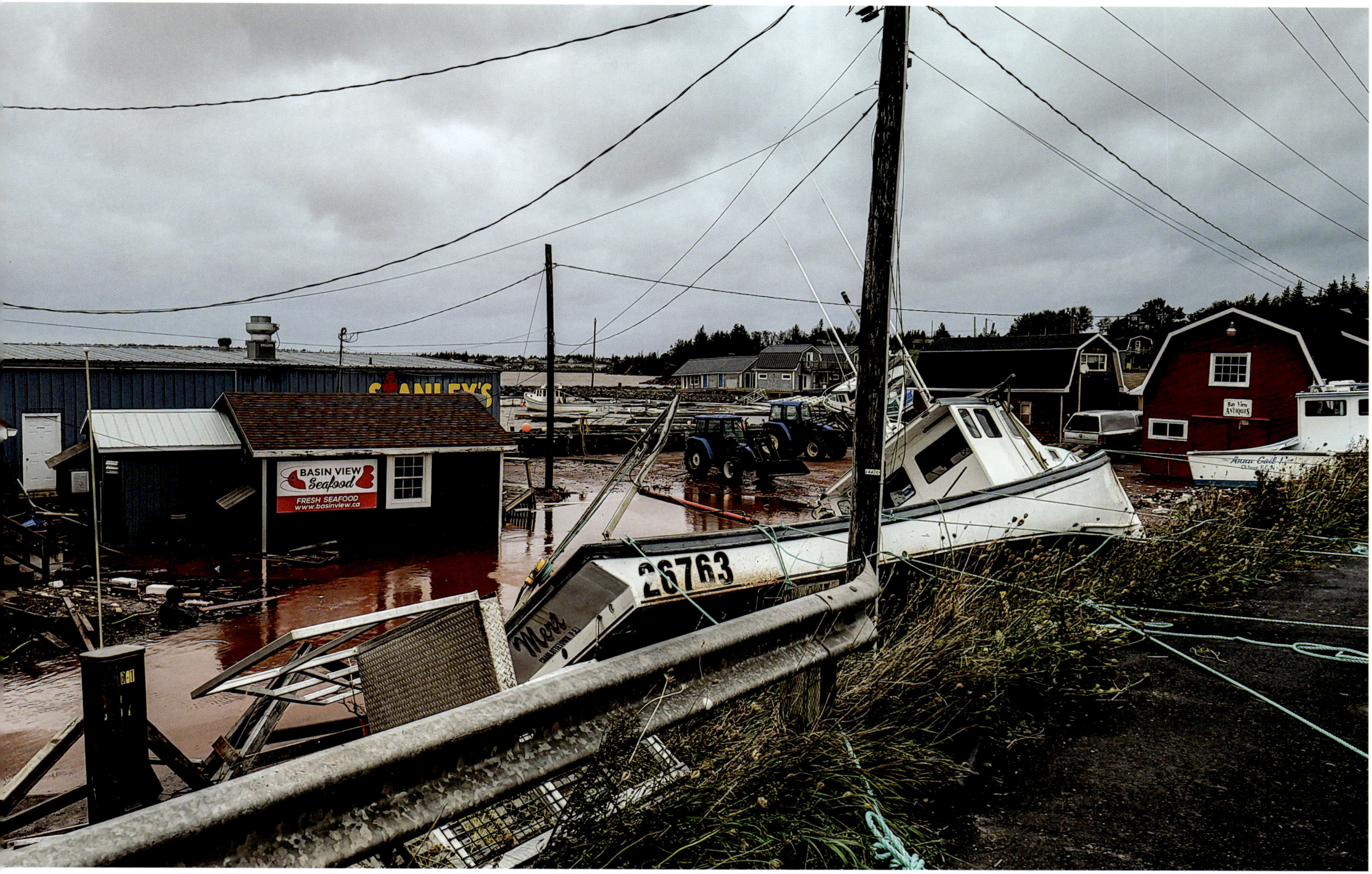

VALFAYE SHAFFER

C A R I N   M A K U Z

SEPT. 26 — The conditions for stargazing have never been finer.

ALEC BEATON

SEPT. 28 — Who even knew woolen blankets draped over a fridge could keep things cold?

SEPT. 29 — The only place in the forest to walk is a sliver of magic inside the entrance, a tiny cathedral of trees somehow left untouched. The smell of freshly snapped wood is everywhere. Now it is my sanctuary, a daily meditation of twenty-five steps. I christen it *Because of Fiona*.

OCT. 2 — Although I have little interest in sports, it's easy to find joy in seeing baseball played on a local field, the community taking a break from despair, making use of things still standing.

OCT. 4 — While harvesting celery, two giant pot-bellied spiders emerge, which would normally terrify me; instead I simply wonder how it was for them.

MARILYN CAREY

One Day at a Time

September 24, 2022

It's 9 AM, and there's no possibility of tea or coffee. The power has been off for almost nine hours, and I don't expect it'll be back on for another eight hours—maybe not even today. That's just a hunch since I can't access the internet and my smartphone battery is almost completely drained, with the words *No Service* flickering on and off the touchscreen. Without internet, I can't follow the government directive to monitor *Hurricanes.ca* for weather updates.

I'm in Charlottetown, miles from the north shore where Hurricane Fiona is pummeling properties and battering the shoreline. It's gusty here, with shingles blowing down the street; our neighbour's attic curtain whipping ghostlike in the space between our home and theirs, their window pane unmoving on the leaf-splattered lawn. With no stove, microwave, or kettle to prepare food, my daily routine of breakfast and tea is off. I can't check my email or read today's top stories. Since we have an electric water heater, I won't be taking my morning shower.

My seven-year-old daughter interrupts my thoughts:

"When will the power be back on?"

"Not for a while, honey. Maybe not today."

"Aww, I wanted to finish watching that movie. What *will* we do today?"

When stripped of the routines of modern life, there's a sense of anxiety at first, a feeling of being cut off from friends and family, from the daily news. And yet, here it is—the day I've been waiting for, the day when the busyness of life stops, when I have a day just *to be*.

September 25, 2022

Writing by candlelight. Nearly forty-eight hours without electricity. The city is abuzz even without traffic lights—headlights whizzing down dark streets, chainsaws droning on in the distance, fire truck sirens winding up and down, generators churning. It is a city unable to sleep, uncertain about the future, wide-eyed at the devastation, taking in the altered landscape.

I drift off to sleep, grateful for my neighbours and friends, my muscles yearning for rest after a day spent raking, collecting branches and leaves, cooking on the camp stove.

September 29, 2022

Friday morning, 6 AM. Darkness to the west, into our neighborhood where heavy trunks lie cradled by low-hanging power lines, other trees cut off or up-rooted in yards or barricading the road, waiting to be cut and carried away. "The trees we planted when each of our children were born are down, blocking our driveway," I heard one caller say through the crackle of our wind-up radio.

To the east, a main thoroughfare, yellow street lights are glowing. I can see traffic and rooftops where I used to see trees, leaves dancing in the breeze. Fences now flattened; so many open spaces.

We mourn with our neighbours. We use their propane stove. The children drag branches cut from the ninety-foot elm tree that fell gracefully into the cemetery behind our houses, its thick core prostrated between two historic gravestones, miraculously leaving both unscathed. Tomorrow, we'll rebuild the fence.

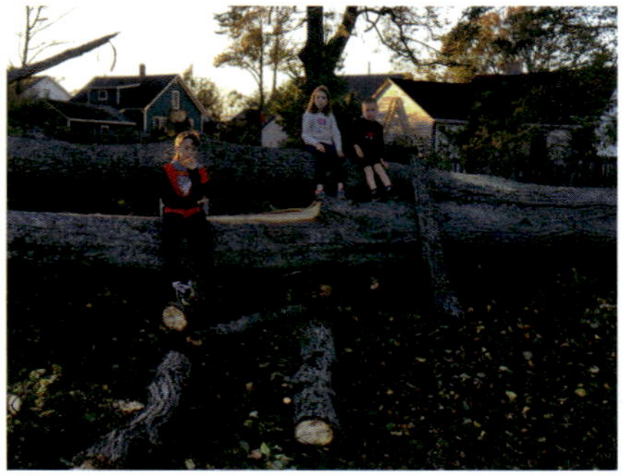

At midnight tonight, it will be a full week without electricity. It's been as if the uprooting of the "old ancestral trees, planted and tended by hands long dead,"[1] has thrown us back to an earlier time, a time without the incessant need to plug in—a time filled with gratitude, a time when wanting waited. Living without refrigeration, our family has had camp stove sour milk pancakes with blueberry sauce made quickly before the blueberries begin to rot. We've had quesadillas crisped on cast iron pans, hot dogs roasted over the fire, hot turkey soup heated under the stars.

I can't wait to take a hot shower—but did you know that washing your hair under the cold tap is surprisingly invigorating? It brings you into the present in a way that the comfort of hot water cannot.

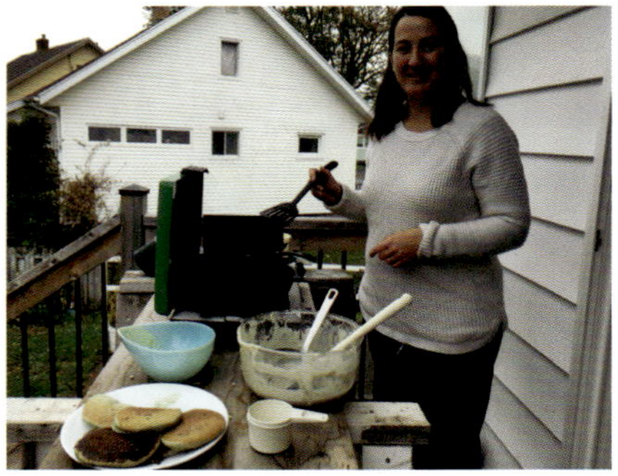

[1] Montgomery, L. M., *Emily's Quest*, Chapter 2.

MICHAEL PENDERGAST

The Great Tree Has Fallen

The great tree has fallen
Its grandeur forgotten
The loot of its branches
Has stolen its wonder
It's taken the gutter
the drainpipe, the soffit
The gate of the garden
Is crushed somewhere under

Leave us in silence
You crows on the rooftop
Complaining your perch
Is no longer so lofty
Where there was a shadow
A presence is hollowed
The bark of the tree
In the earth has been swallowed

This friend of the air
Was brought to its knees
First leaning, it stumbled
Its leafy limbs fumbled
The foliage trembled
Its roots in a jumble
Oh see, what a sin
Sighs the treacherous wind

The sidewalk has lifted
The giant has fallen
The houses are naked
Perspective has shifted
And I, for one
Am saddened and troubled
Rings of the age
Have been lost in the rubble

JUSTIN SIMARD

Fall

Autumn is usually a feast for the eyes,
as leaves turn to vibrant
yellow, orange, red, or purple.

Instead, this year, the trees
drape themselves
in mourning shades
of shriveled brown
as they hold their fallen friends,
weighty in their branches
pressing on their trunks

We mourn with you, trees
Fiona has left a scar on the forest
and robbed us of Autumn,

leaving
only
Fall.

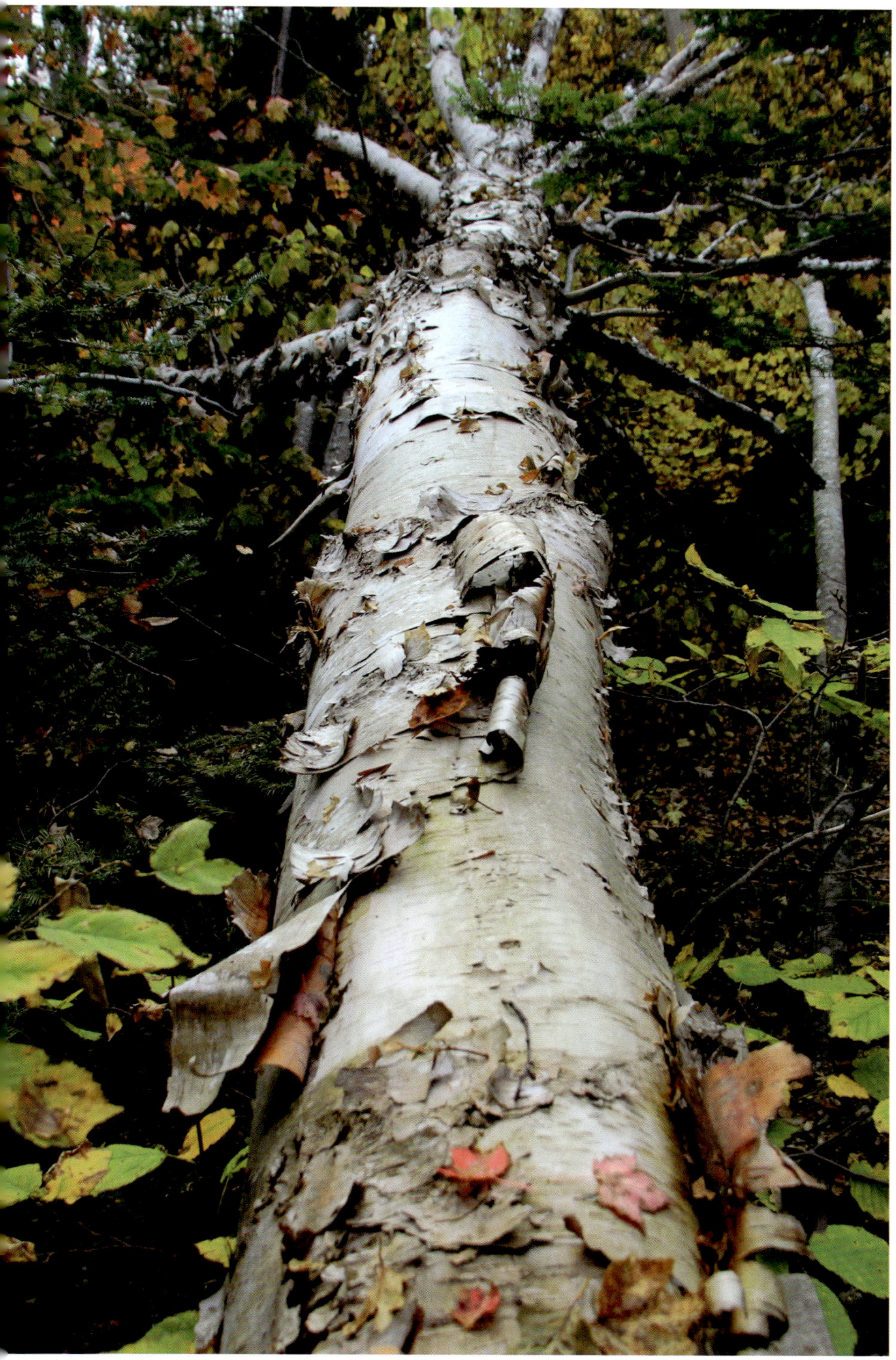

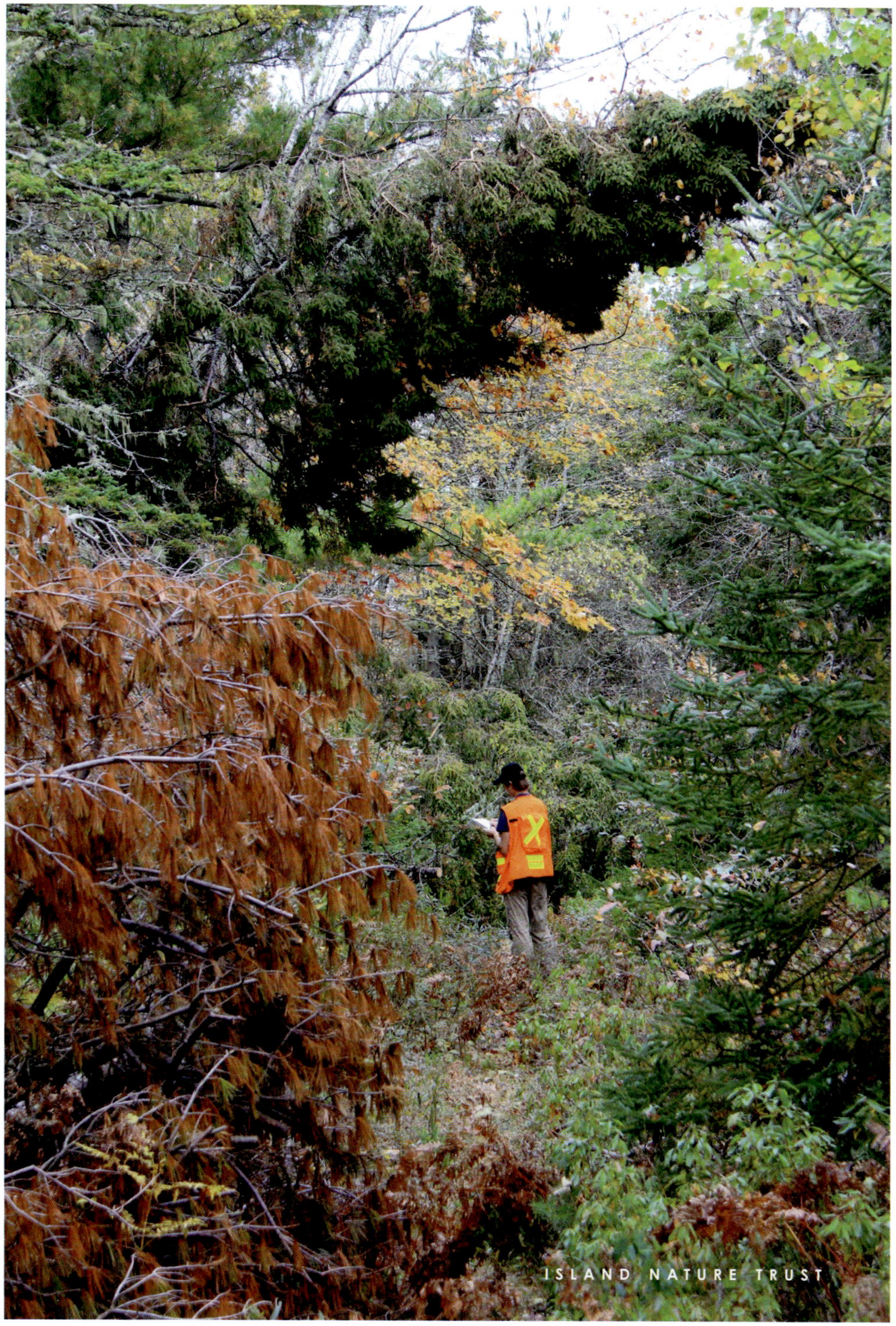

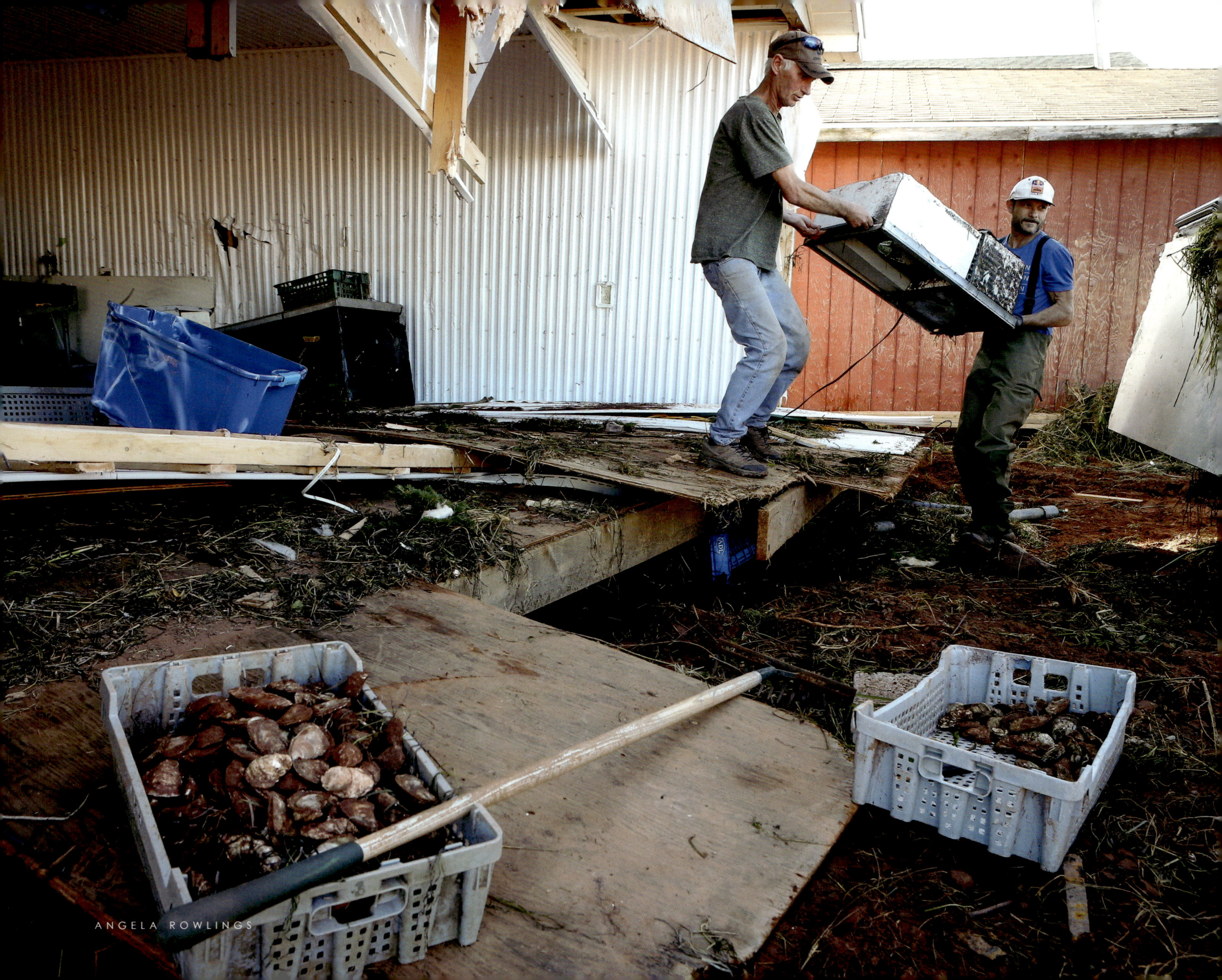

ANGELA ROWLINGS

FARMERS AND FISHERS

ARTISTS AND DREAMERS

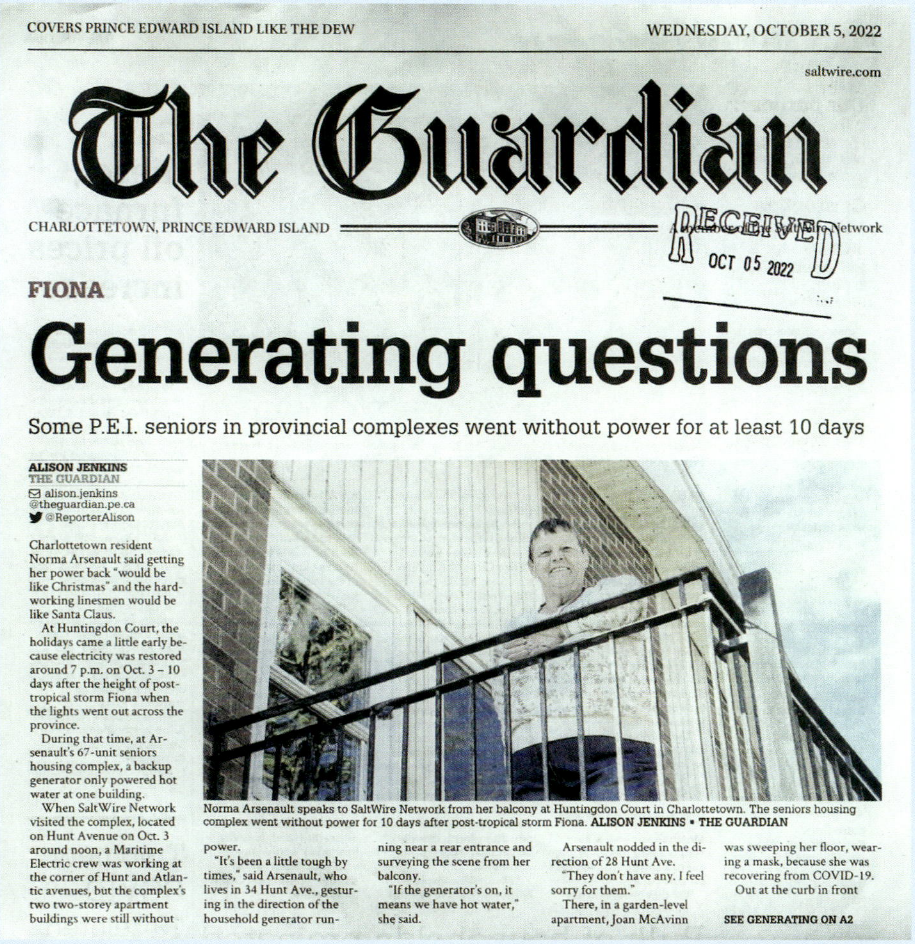

COVERS PRINCE EDWARD ISLAND LIKE THE DEW

WEDNESDAY, OCTOBER 5, 2022

saltwire.com

The Guardian

CHARLOTTETOWN, PRINCE EDWARD ISLAND

RECEIVED OCT 05 2022

FIONA

Generating questions

Some P.E.I. seniors in provincial complexes went without power for at least 10 days

ALISON JENKINS
THE GUARDIAN
alison.jenkins@theguardian.pe.ca
@ReporterAlison

Charlottetown resident Norma Arsenault said getting her power back "would be like Christmas" and the hard-working linesmen would be like Santa Claus.

At Huntingdon Court, the holidays came a little early because electricity was restored around 7 p.m. on Oct. 3 – 10 days after the height of post-tropical storm Fiona when the lights went out across the province.

During that time, at Arsenault's 67-unit seniors housing complex, a backup generator only powered hot water at one building.

When SaltWire Network visited the complex, located on Hunt Avenue on Oct. 3 around noon, a Maritime Electric crew was working at the corner of Hunt and Atlantic avenues, but the complex's two two-storey apartment buildings were still without power.

"It's been a little tough by times," said Arsenault, who lives in 34 Hunt Ave., gesturing in the direction of the household generator running near a rear entrance and surveying the scene from her balcony.

"If the generator's on, it means we have hot water," she said.

Arsenault nodded in the direction of 28 Hunt Ave.

"They don't have any. I feel sorry for them."

There, in a garden-level apartment, Joan McAvinn was sweeping her floor, wearing a mask, because she was recovering from COVID-19.

Out at the curb in front

Norma Arsenault speaks to SaltWire Network from her balcony at Huntingdon Court in Charlottetown. The seniors housing complex went without power for 10 days after post-tropical storm Fiona. **ALISON JENKINS • THE GUARDIAN**

SEE GENERATING ON A2

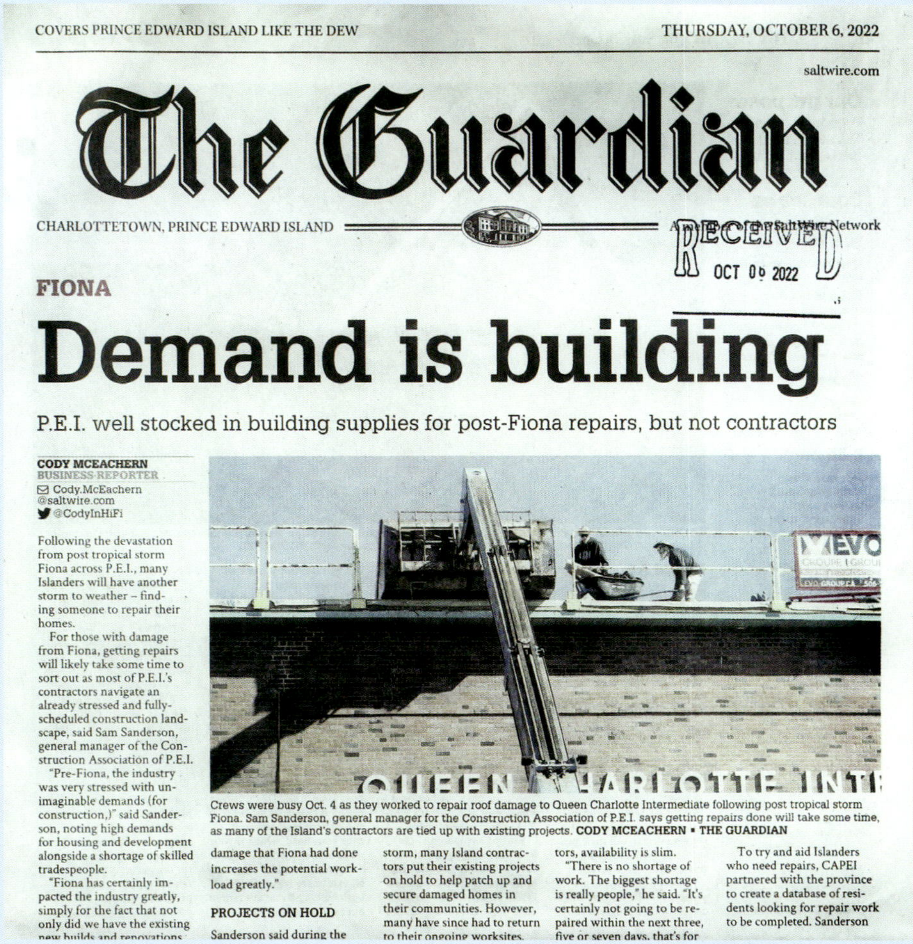

COVERS PRINCE EDWARD ISLAND LIKE THE DEW

THURSDAY, OCTOBER 6, 2022

saltwire.com

The Guardian

CHARLOTTETOWN, PRINCE EDWARD ISLAND

RECEIVED OCT 06 2022

FIONA

Demand is building

P.E.I. well stocked in building supplies for post-Fiona repairs, but not contractors

CODY MCEACHERN
BUSINESS REPORTER
Cody.McEachern@saltwire.com
@CodyInHiFi

Following the devastation from post tropical storm Fiona across P.E.I., many Islanders will have another storm to weather – finding someone to repair their homes.

For those with damage from Fiona, getting repairs will likely take some time to sort out as most of P.E.I.'s contractors navigate an already stressed and fully-scheduled construction landscape, said Sam Sanderson, general manager of the Construction Association of P.E.I.

"Pre-Fiona, the industry was very stressed with unimaginable demands (for construction,)" said Sanderson, noting high demands for housing and development alongside a shortage of skilled tradespeople.

"Fiona has certainly impacted the industry greatly, simply for the fact that not only did we have the existing new builds and renovations damage that Fiona had done increases the potential workload greatly."

PROJECTS ON HOLD

Sanderson said during the storm, many Island contractors put their existing projects on hold to help patch up and secure damaged homes in their communities. However, many have since had to return to their ongoing worksites. tors, availability is slim.

"There is no shortage of work. The biggest shortage is really people," he said. "It's certainly not going to be repaired within the next three, five or seven days, that's for

To try and aid Islanders who need repairs, CAPEI partnered with the province to create a database of residents looking for repair work to be completed. Sanderson

Crews were busy Oct. 4 as they worked to repair roof damage to Queen Charlotte Intermediate following post tropical storm Fiona. Sam Sanderson, general manager for the Construction Association of P.E.I. says getting repairs done will take some time, as many of the Island's contractors are tied up with existing projects. **CODY MCEACHERN • THE GUARDIAN**

BONNIE FLAVELL

Seeds of Change

THERE'S SOMETHING ABOUT TREES. Children run to them, grown-ups cherish them. Much like us, trees are rooted in family and grow stronger in diverse communities. We have always shared an intimate connection, rooted in respect. Have we parted ways? Trees give so much and ask so little; it is a testament of their forgiving nature that they remain steadfast and true. Each time we return to them, they welcome us into the heart of the wood, in the hopes that we will lean in, reach out, and reconnect.

The forest is a spiritual place: one needs only to step inside to feel its healing touch of ecological harmony. It embodies wisdom, awakens our senses, and offers clarity in a world of chaos. Trees store the carbon emissions that we create and give us fresh air. Yet, across the globe, our old-growth forests continue to be sacrificed. These entities of enlightenment hold life on Mother Earth together. Here on Prince Edward Island, we are especially vulnerable. Without trees, our Island would be as fragile as our sand dunes.

Now, just three years since Dorian struck, we are dealing with the aftermath of yet another hurricane. Fiona shook us to our core. She displaced people, birds, animals, and more little creatures than we will ever know. In her fit of fury, she took lives, devastated shorelines, destroyed homes, and uprooted trees by the thousands, compounding our carbon footprint.

Fall is the best time of year to connect with nature, but many of the trails in our provincial parks were obliterated, and our forests have now become danger zones. Despite this stark situation, when the sky cleared, one person after another reached out to offer a helping hand. One tree at a time, we set out to work on the recovery. First responders, Maritime Electric, the military, and crews from across Canada and beyond worked around the clock. Everywhere we looked, people were helping people.

The living memories of the fallen will be honoured in works of art and in our hearts. Though it grieves us to lose so many, even in death, trees will give life.

Forests can heal themselves, but it will take time. In a fraction of the time it will take the forests to regenerate, we will have cleared our properties and rebuilt our homes. Will we have time to rebuild our relationship with nature and stop this vicious cycle before the next storm?

Trees are the guardians of Mother Earth. They must be protected. Without trees, what life would there be? Now more than ever, we need to plant the seeds of change and help them grow.

Fiona's Harvest

Fiona, you blowhard,
you wreaked devastation,
ravaged thousands of trees
and laid waste to the dunes.
Things that stood centuries
lie broken, in ruins,
all in one fearful night.

It will take generations
to replace your cruel harvest,
as the sea level rises,
and the oceans keep warming,
driving hurricanes roaring,
farther north up the coast.

Fiona, you battered
this fragile small island,
its forests and sandstone,
red cliffs, farms, and harbours.
Ephemeral beauty
that you chewed like a bone,
all in one tearful night.

It will take generations
to replace your bleak harvest.
But the farmers and fishers
and the artists and dreamers
will cling to this island, like
marram grass to its dunes.

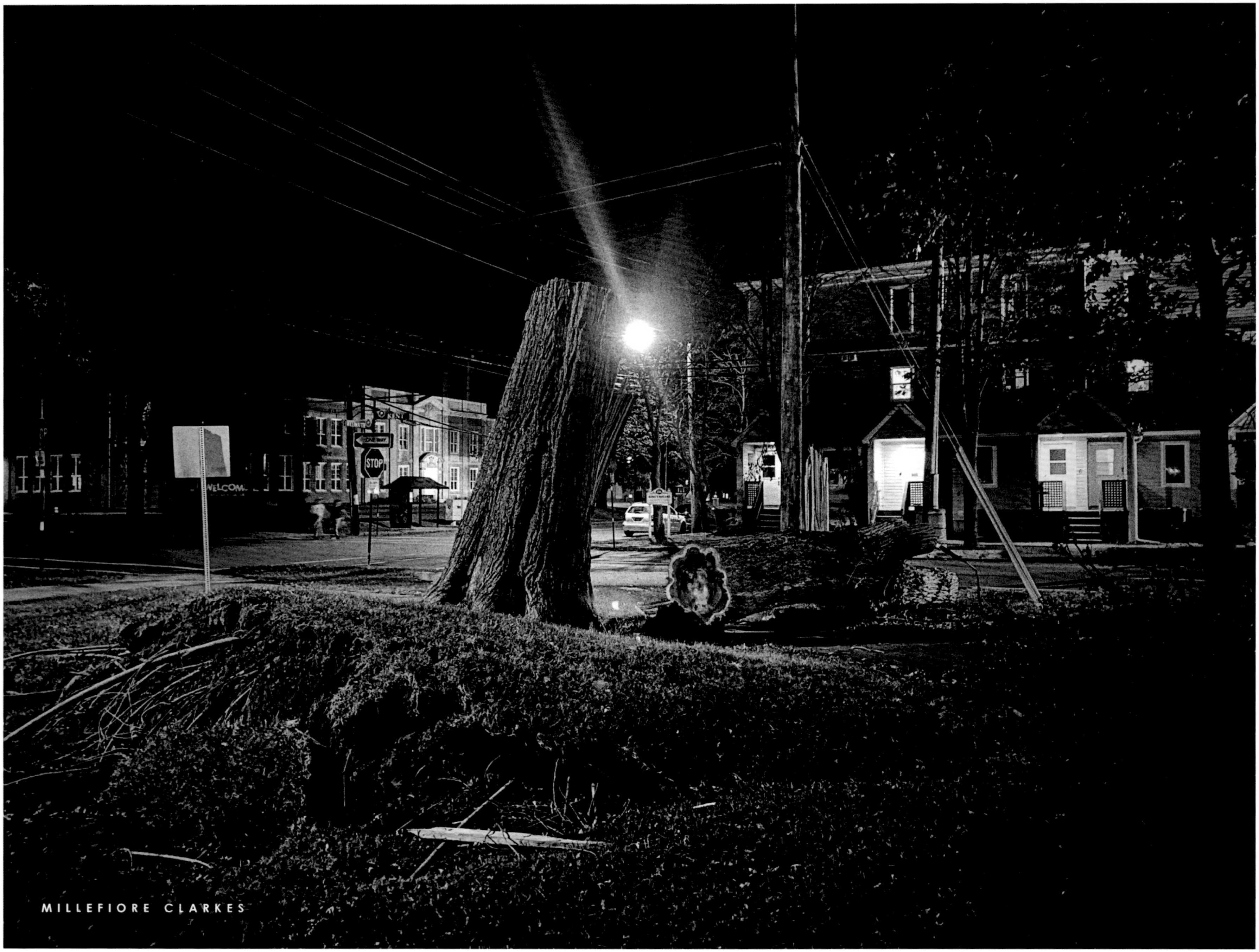

MILLEFIORE CLARKES

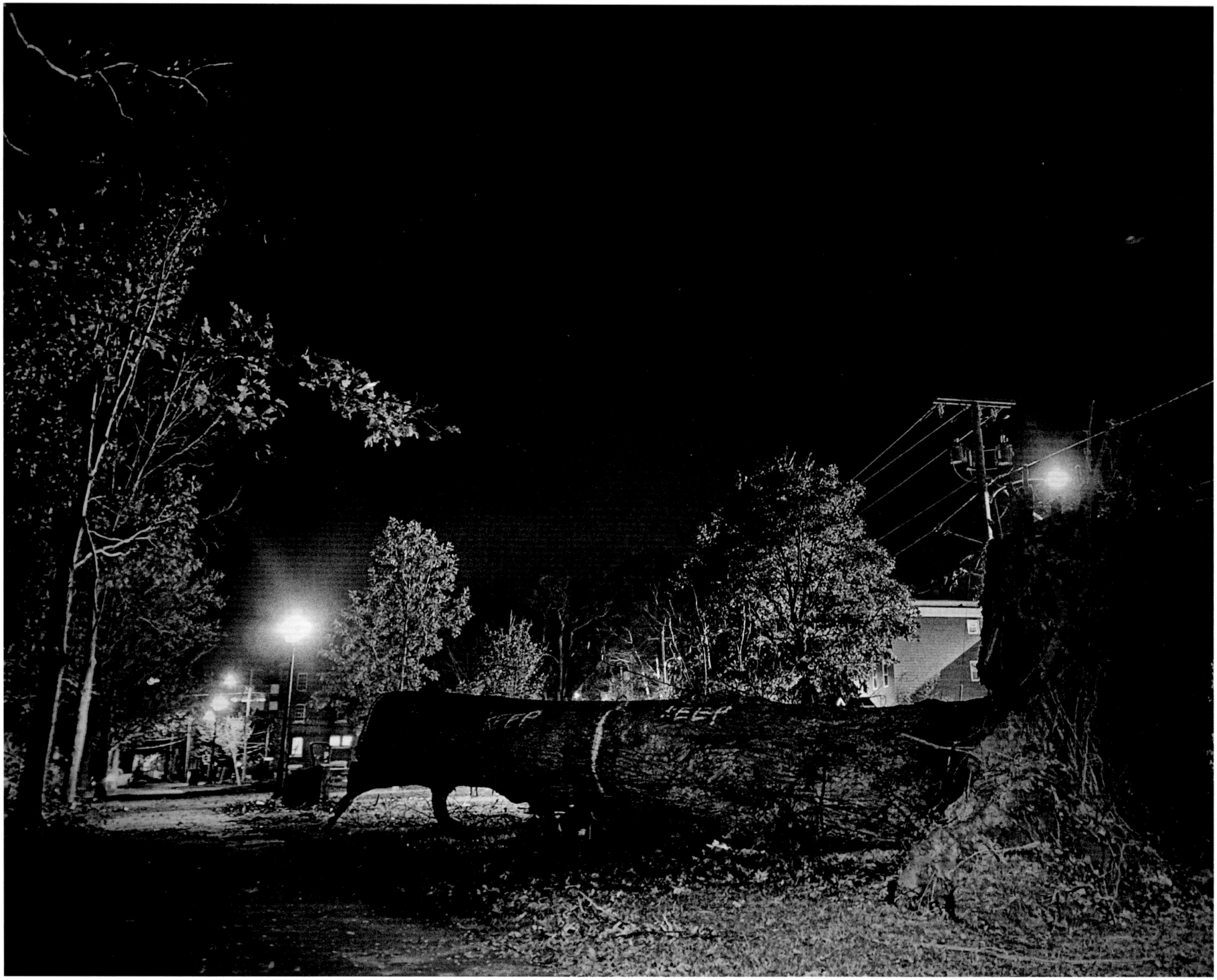
KEEP
KEEP

Strange Day

What a strange day it was today.
The air, breathless, and ripe with anticipation for something
uneasy.
Hot. Weirdly hot for October (in Canada).
The streets almost empty. Those that wandered walked subdued.
And all the giant trees lying down.
And the festering hoard of flies in the compost bin
(feasting on a chicken carcass discarded when our island went dark
for days and days).

And the 500 dead whales in New Zealand.

I no longer know how to think about the future.

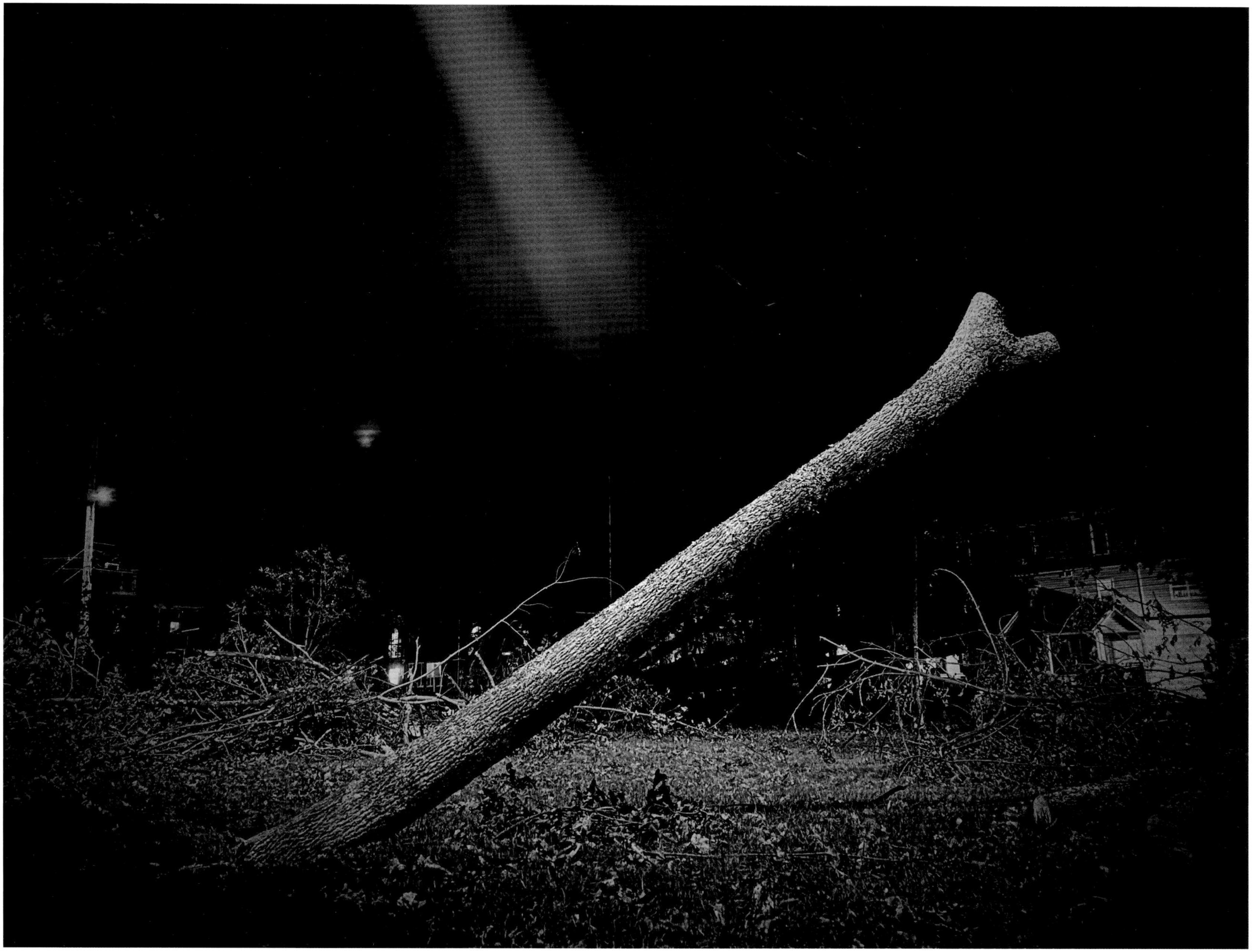

JANE AFFLECK

Wrack Line

The day after, the estuary is a purple vein pumping into the strait.
You go to the shore to walk the wrack line, as you often do,
to read that rising–falling chart of the tide,
the heartbeat of the moon inscribed by the ocean.
The day after, the wrack line is off any chart you've ever seen.
The sea has sunk its teeth into the shoreline,
bitten off more than it could chew
from the sodden cake of this island, scattered sticky crumbs
of clay beneath the cliff, rivulets of rusty red soaking the sand.
The waves, drunk on power, disgorged this storm-surge spew,
indigestible matter, this wrack, a line crossed.

Great blue heron—lone
survivor, turns head and neck
to ask a question.

You scrutinize the wrack, stock-still, heart an iceberg in your chest.
Tangled skeins of black and green seaweed, wider than rows of mown hay.
Thick nests of cattail reeds, not for eggs but carcasses:
the salt-scored breasts of an eider duck, a double-crested cormorant.
Wind-ripped plastic bags; mud-stained water bottles; crumbling buoys.
The stringer ripped from a staircase, stripped of paint, stepless.
Branches thick as arms, stacks of horizontal trees,
resembling shelter but offering no refuge, only menace.
Ripe apples, dashed onto the mud, bright red on red, their bruises
release a sweet and dissonant scent.

That bird, this plastic
lid: your eyes ken which one
they'll see in twenty years.

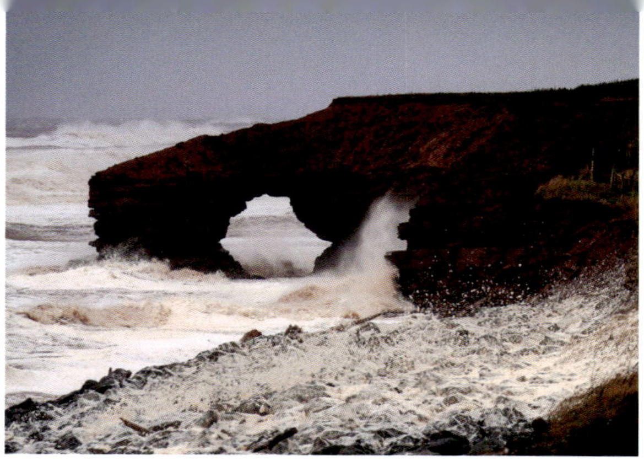

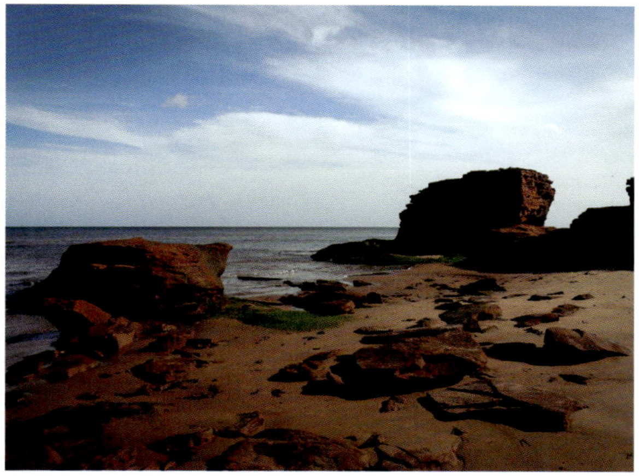

You comprehend anew that wrack is aptly named—
a word from centuries ago, when canvas sails sought to harness wind;
an index for the wreck of baffled ships, their lading spilled and beached.
Of damage and destruction, leading to collapse and ruin.
The shift from wind to coal and diesel, the inauguration
of a timeline toward a tipping point; a line drawn
on sand but standing as an augury of higher lines to come.
Here, these lines of wrack hold liminal space,
neither land nor water: like the heart's break
                    between contractions,
an interstice when the tide could turn either way.

    Remain in arrest,
    or resume the pace,
    in spite of the ache.

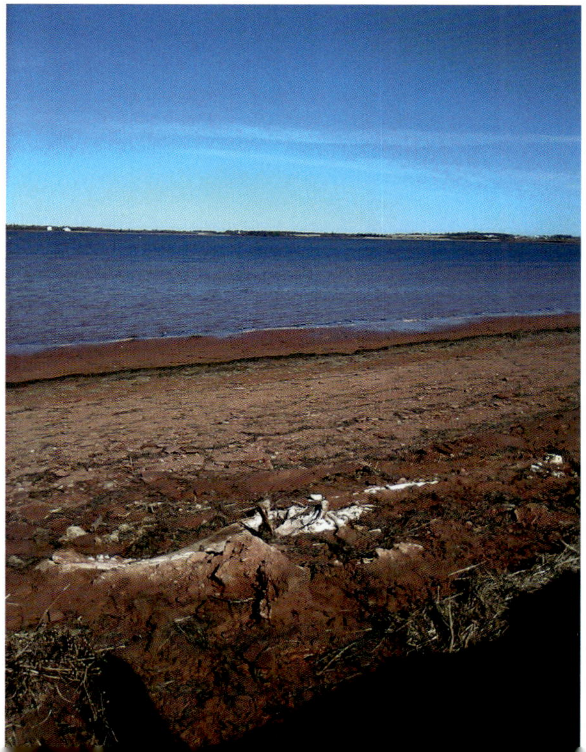

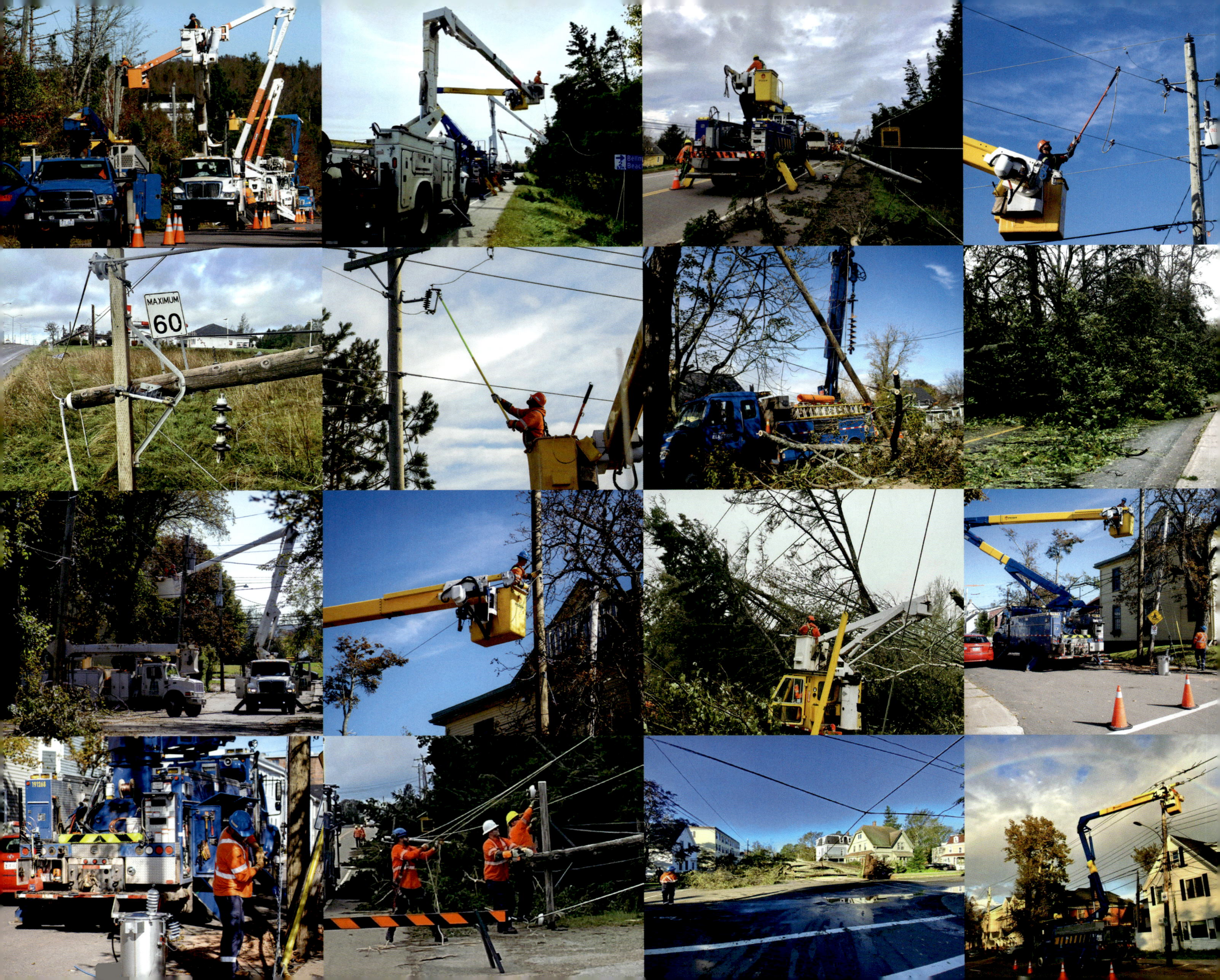
MAXIMUM
60

KIM GRIFFIN

Twenty-One Days

LOOKING BACK, I RECALL THE SADNESS AND EERIE FEELING as I ventured out in Fiona seeing the hundreds of trees down across the streets and hearing the same comments from all across the Island about the devastation and loss—not only of power, but damage to roads, bridges, homes reaching tip to tip across the Island.

We prepared for the storm as best we could. A team of people across Maritime Electric had been mobilized for days, reaching additional crews and personnel in advance of the storm and activating teams to mobilize when we needed them. As the wind started to howl on Friday night, and as Fiona hit landfall in the region, we braced ourselves. With apprehension and measures in place for the response, we were on the ready to see what Mother Nature would deliver to us.

Our first few days and hours after the storm were intense. It was immediately apparent that Maritime Electric had never dealt with anything of this magnitude, but the adrenaline kicked in, and our teams ventured out. We knew what we had to do, and it became a race against time—all with a focus on safety. Assessment and action began all across the Island, as trees were removed from roads and lines. It wasn't chaotic; it was focused, deliberate action to restore power and keep people safe in the process.

Our biggest challenge: time. We needed time to remove the trees and gain access across the roads to even begin to get to our customers and begin the power restoration. Helicopter inspections and drones captured epic visuals of the damage and destruction across Prince Edward Island. With each restoration job, our crews responded as quickly and as safely as possible. Some days, each job seemed worse as new destruction and impact of the downed trees was found everywhere. The majestic trees all across our communities that cascaded along our lines were left in Fiona's wake, instantly removing the power from people's homes and businesses. Boots on the ground all across our Island focused on restoring power.

Over the hours and days that followed, the help poured in. Luckily, with advance planning and mutual aid agreements, the Fiona restoration was activated, with over 260 crews and hundreds of employees working at the height of the storm; it had become the largest storm response in Maritime Electric's history. Support from our parent company, Fortis Inc. in Newfoundland, sister utilities from Newfoundland Power, FortisBC, Fortis Alberta, Fortis Ontario and Central Hudson began to arrive. These crews worked together with Island crews and contractors, shoulder to shoulder working to restore power to our customers. Across the Island in hotels, crews were housed in places without power for many days. Lines were reenergized across the province piece by piece when trees were removed. Crews often commented about leaving and returning in the dark, day after day, as power was systemically and meticulously restored. Many of our employees left their own homes with damage to work the aftermath and return power to customers. All were thankful for family and friends who assisted in taking care and helping each other.

The Canadian Armed Forces Task Force PEI worked closely with our team as well as emergency responders and leaders across all areas of government and community groups, with fire departments, police personnel, transportation workers, and front line personnel all reaching out and doing their part to help Islanders. Everyone helped us.

Across the Island, we heard hundreds of stories of Islanders helping each other with generators, tree removal, food, shelter and whatever was needed to help. Our teams worked through the days and through the nights, and many in the community offered help and food and showed kindness to our teams. Kind words, invitations to Thanksgiving dinner for off-Island crews away from their family, biscuits, tea and offerings of all kinds were made to show appreciation to our crews. Offers from local businesses and their equipment flooded in to assist with repairs; personnel and food kept the restoration team powered to do their work safely.

Together with thousands of other people across PEI in Fiona, we managed to safely restore power to every single customer on PEI. Although it took twenty-one days, there was relief when the vast majority of customers were restored in western PEI, but there was still more work to be done. There were some customers who had access issues and individual outages or property damage that our crews were still working with! For weeks after Fiona, into December, crews continued to stabilize the system and clean up. Fiona has had a huge impact on our team, as the biggest storm in our history. Looking ahead, the team is planning for the future, focusing on lessons learned and looking at climate adaptation planning. Maritime Electric is committed to a sustainable future for our customers, and with Fiona in the rear-view mirror, there were some key messages to go back to. Each day, the entire team at Maritime Electric focuses on keeping the power on for customers—today and every day.

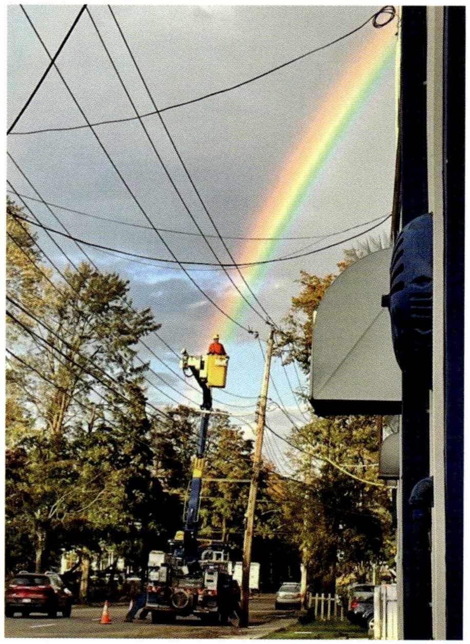

—Kim Griffin

Manager, Sustainability & Corporate Communications, Maritime Electric

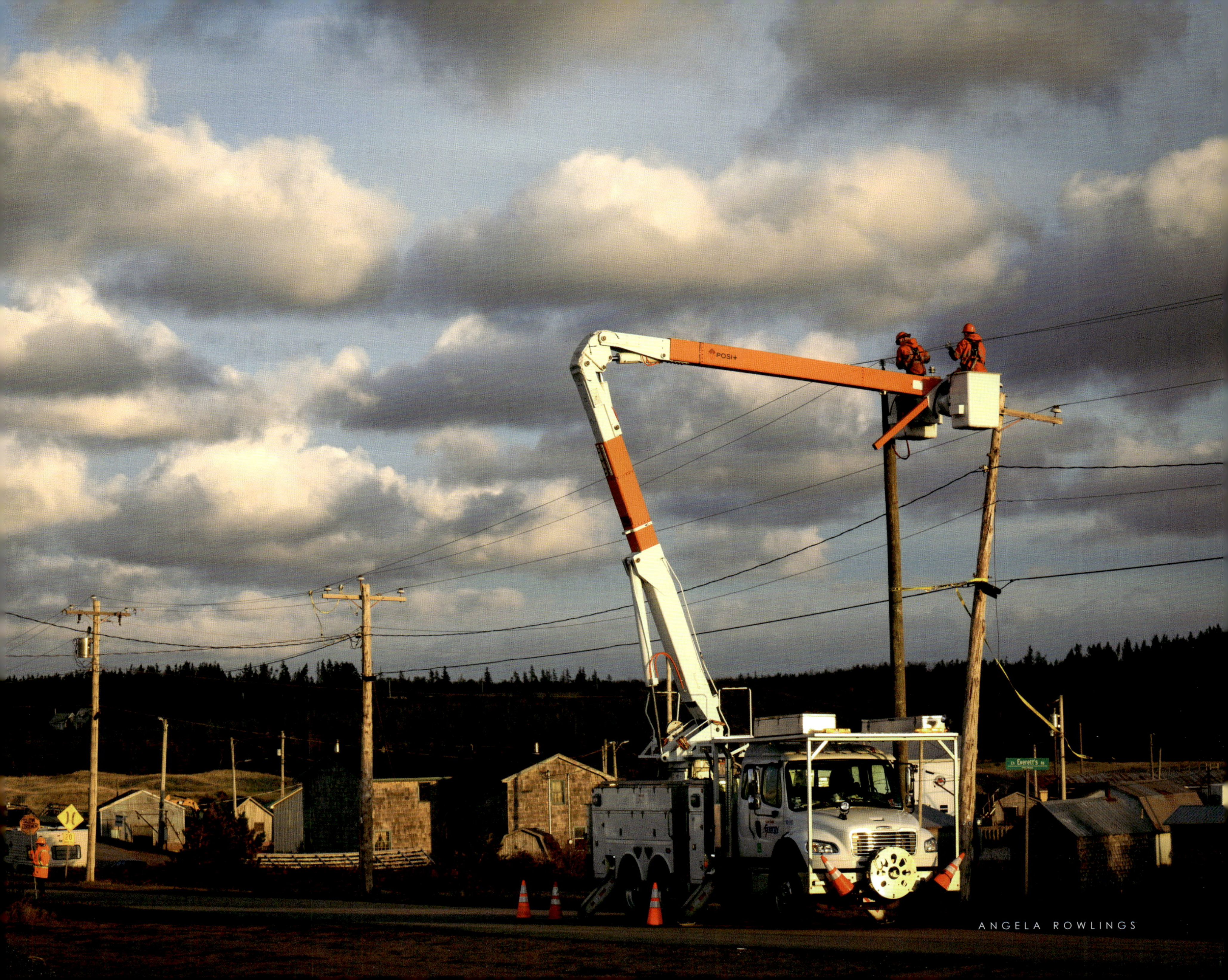

ANGELA ROWLINGS

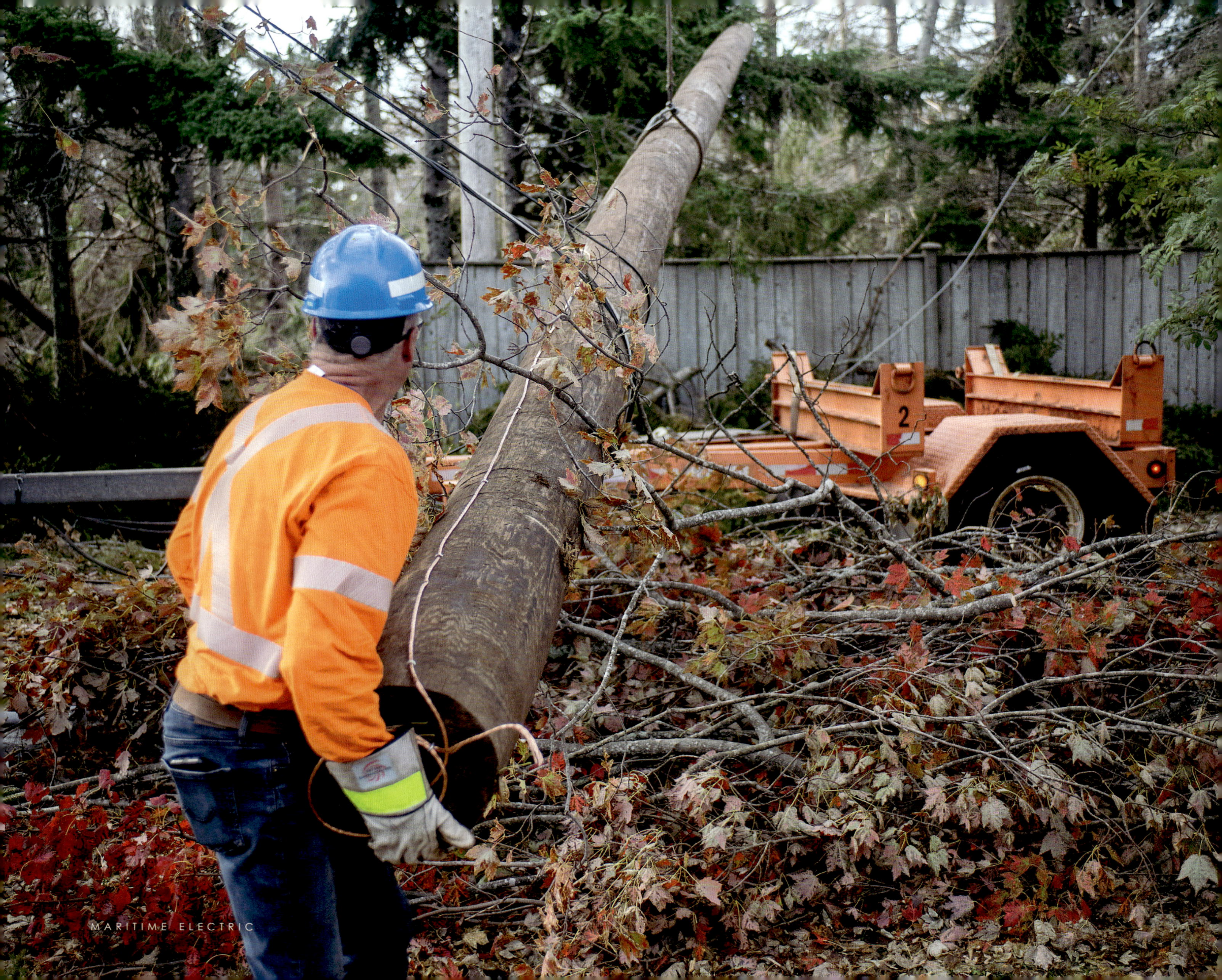
MARITIME ELECTRIC

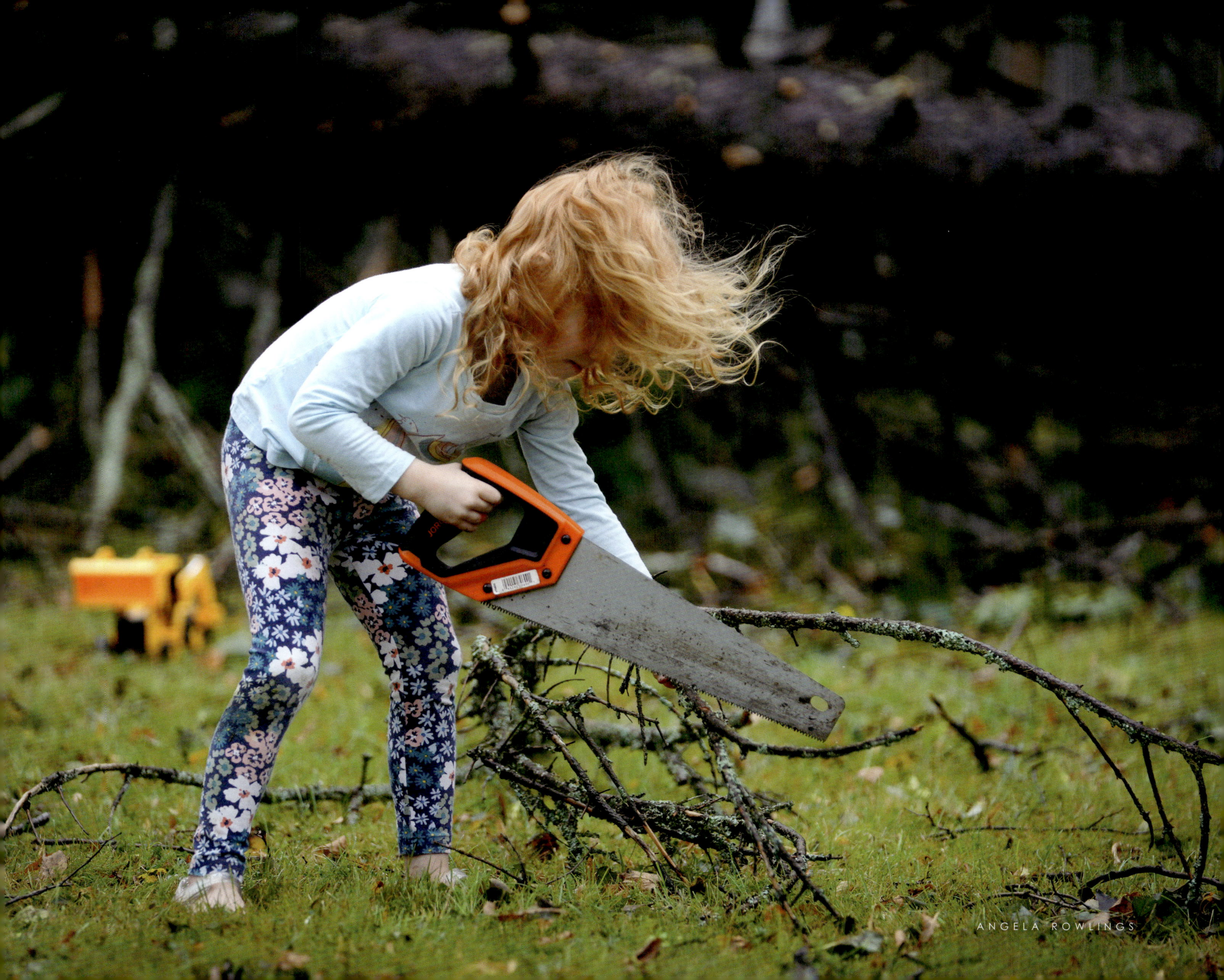

ANGELA ROWLINGS

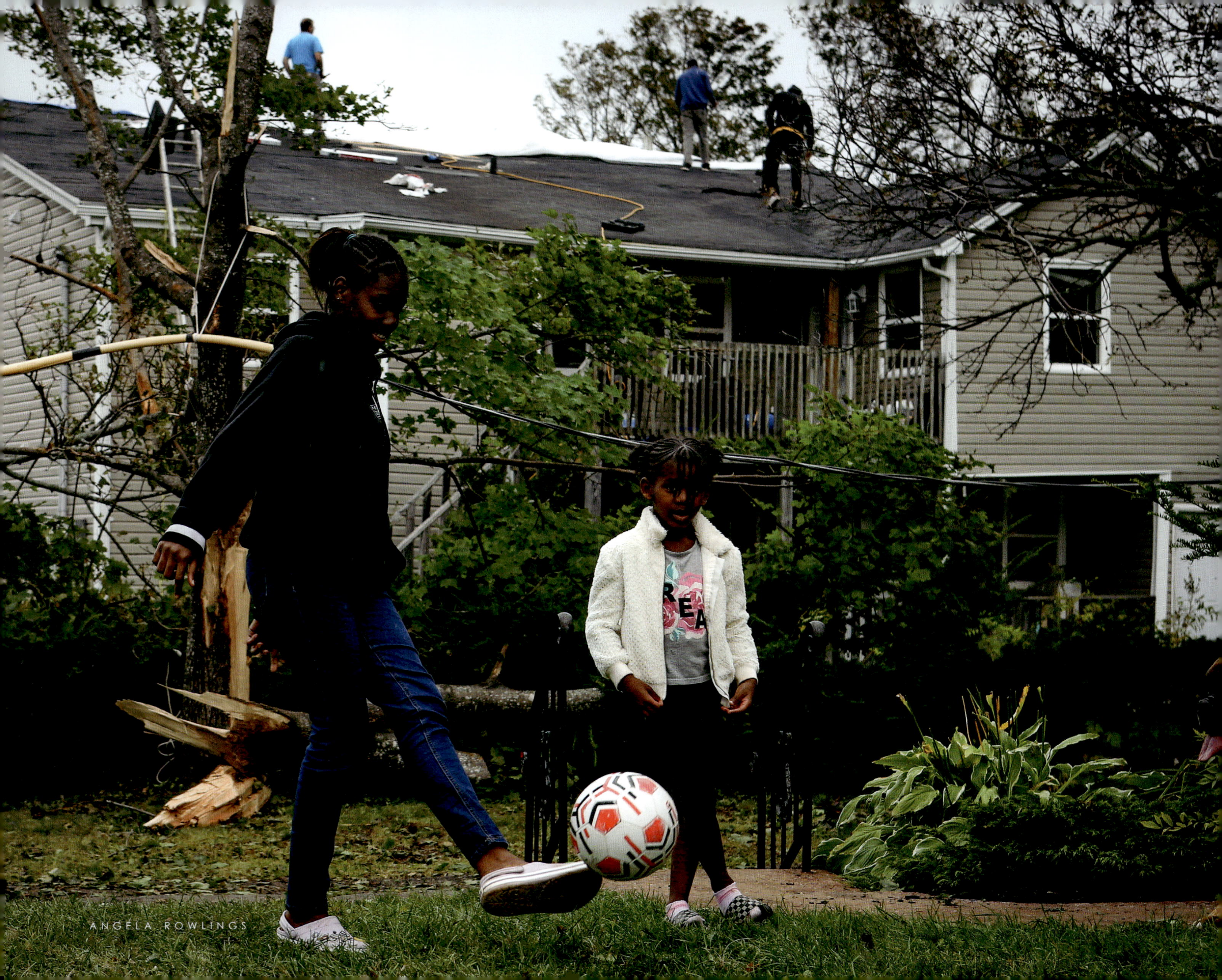

ANGELA ROWLINGS

COVERS PRINCE EDWARD ISLAND LIKE THE DEW FRIDAY, OCTOBER 7, 2022

saltwire.com

The Guardian

CHARLOTTETOWN, PRINCE EDWARD ISLAND A member of the SaltWire Network

FIONA

Creations from chaos

Charlottetown's destroyed trees to be recycled, turned into furniture, art pieces

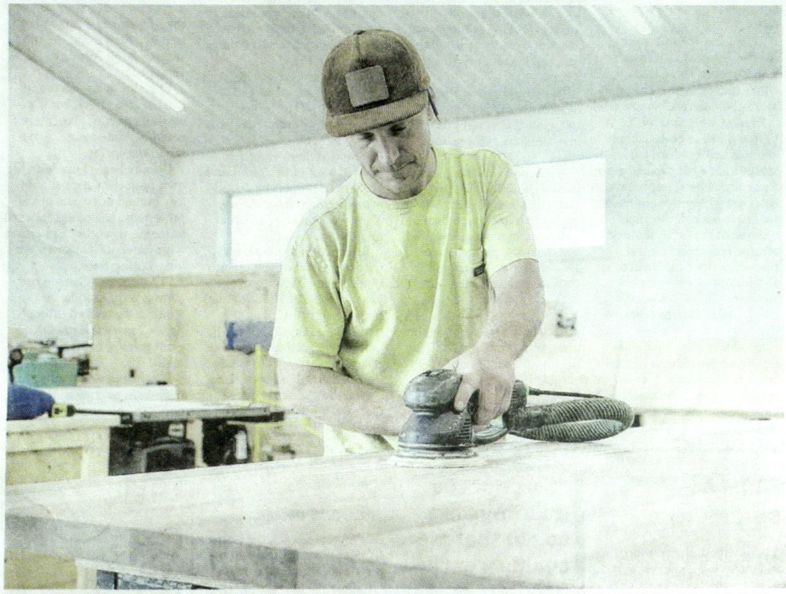

Brodye Chappell, owner of Brodye Chappell Builds, says he's excited to be able to turn Charlottetown's fallen trees into furni-

CODY MCEACHERN
THE GUARDIAN
Cody.McEachern
@saltwire.com
@CodyInHiFi

While post-tropical storm Fiona left many of Charlottetown's trees destroyed, some of the city's decades-old trees are being recycled into artisan fixtures and furniture.

As brush and debris is being collected at a temporary holding lot set up on Park Street, the City of Charlottetown is setting aside large salvageable stumps and logs from many of its older, mature trees for a special project.

"We're trying to find ways we can repurpose these and build some heirloom pieces that are going to last another 200 years," said Brodye Chappell of Brodye Chappell Builds.

Chappell is one of two local furniture builders, the other being Elliot Mallett of Elliot Mallett Woodwork, working with the city to recover and repurpose some fallen trees into unique artisan wood pieces.

Chappell said he and Mallett were asked to join the project in the first few days

JESSICA FRITZ

Chewed Up

Drinking tea, on the stoops
Next to the broken, my heart is heavy
But grateful

What can be said?

Glad for no injuries,
No true damage, but
Uncried tears
From holding strong, standing strong

Will the rivers wash away the rocks
Like the winds did the sands?

In a world mixed up
Whirled around, spat out
For us to digest

To digest.

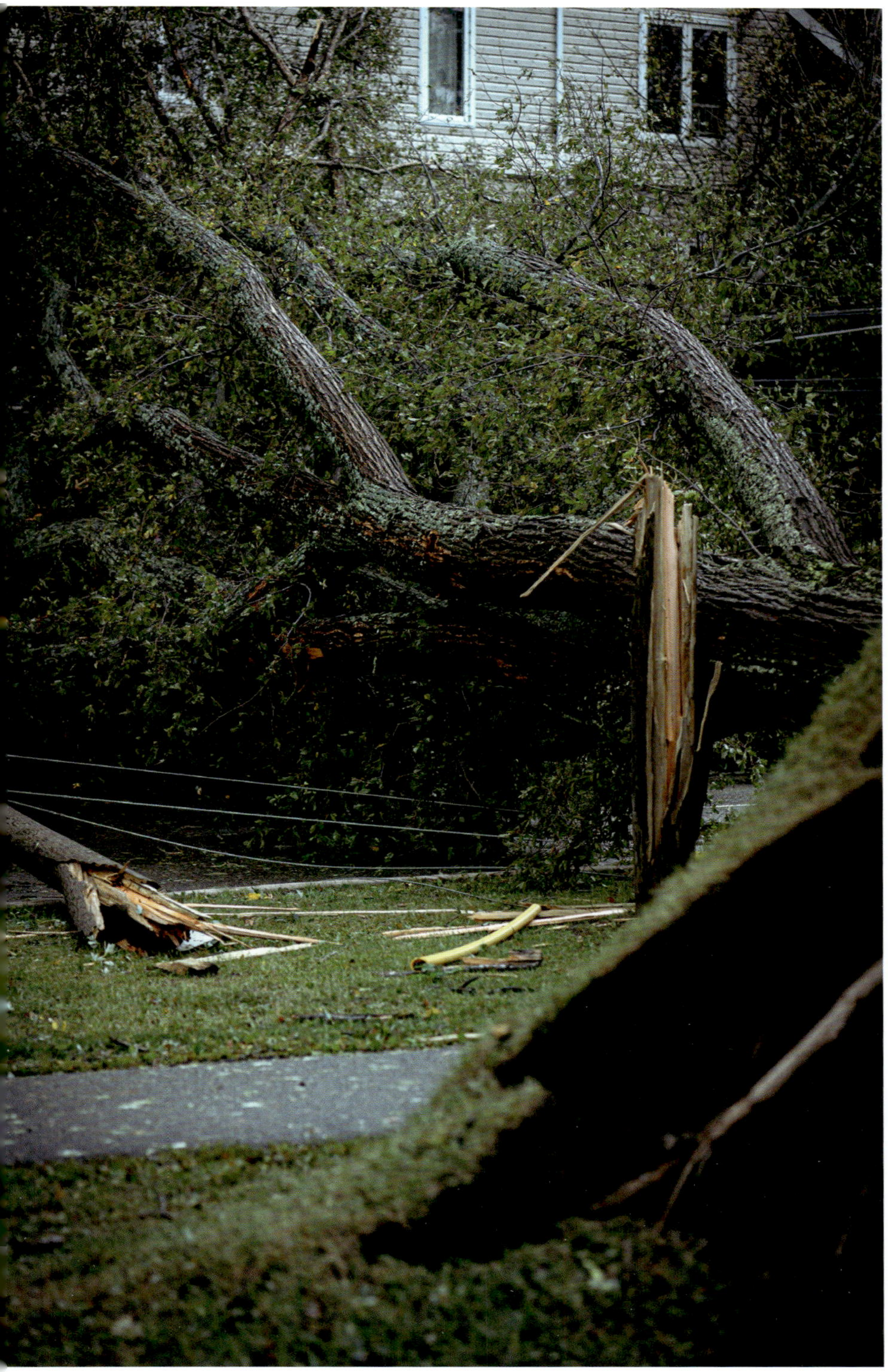

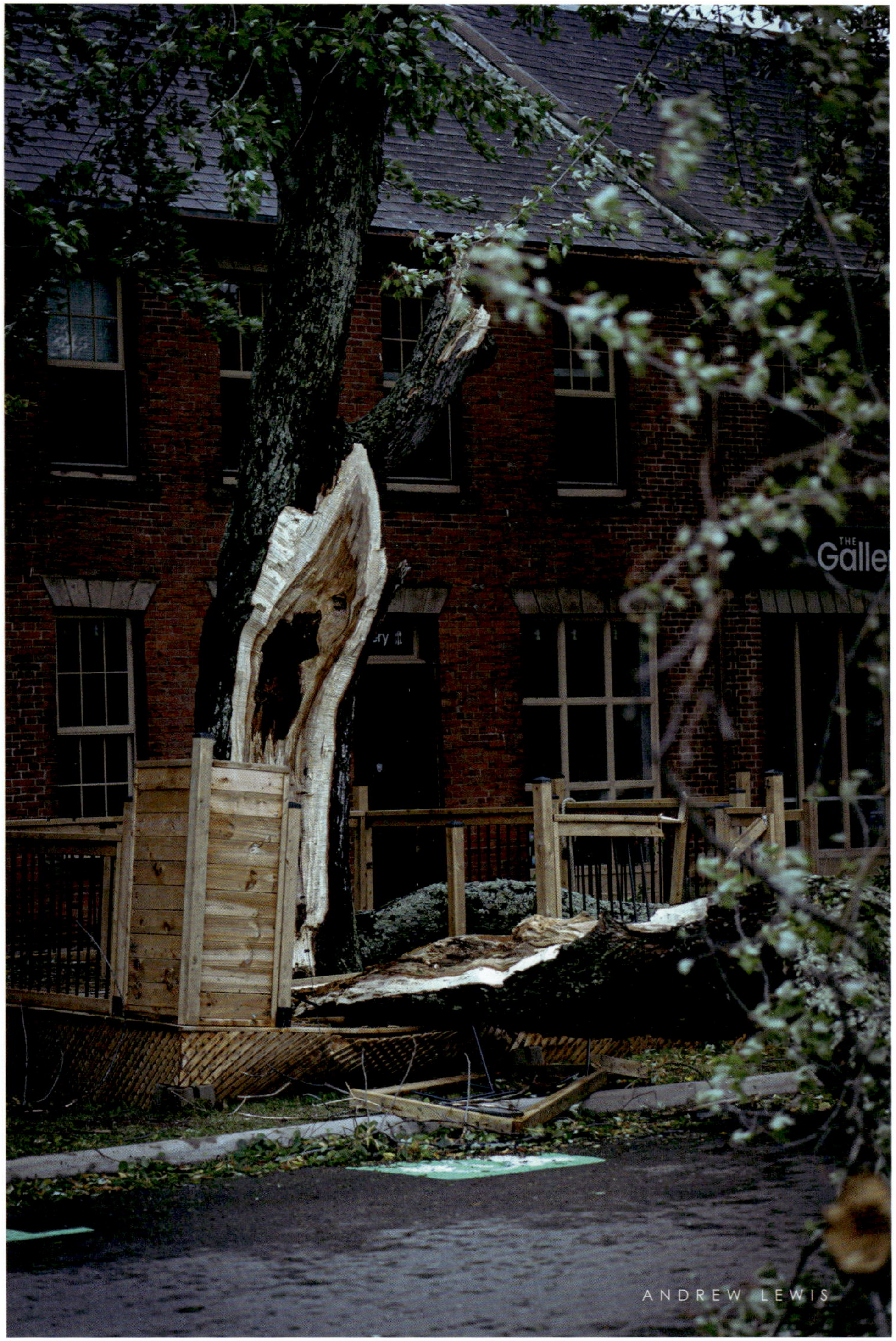

ANDREW LEWIS

OCT. 7 — Power back on. First thing, a load of laundry. My tea towels and assorted whatnots have never been happier. Next thing, a shower.

OCT. 9 — I thought once power was restored all would be normal again, but apparently it's not that simple. We wander slightly trance-like, organizing spice shelves and tea cabinets. Everyone is still shook up.

OCT. 10 — Grateful for the smell of supper cooking . . . on the stove.

OCT. 12 — Earliest spring flowers in magnificent bloom. Are they as confused as the rest of us?

PHOTOS BY BECCA GRIFFIN AND ALANA SPRAGUE

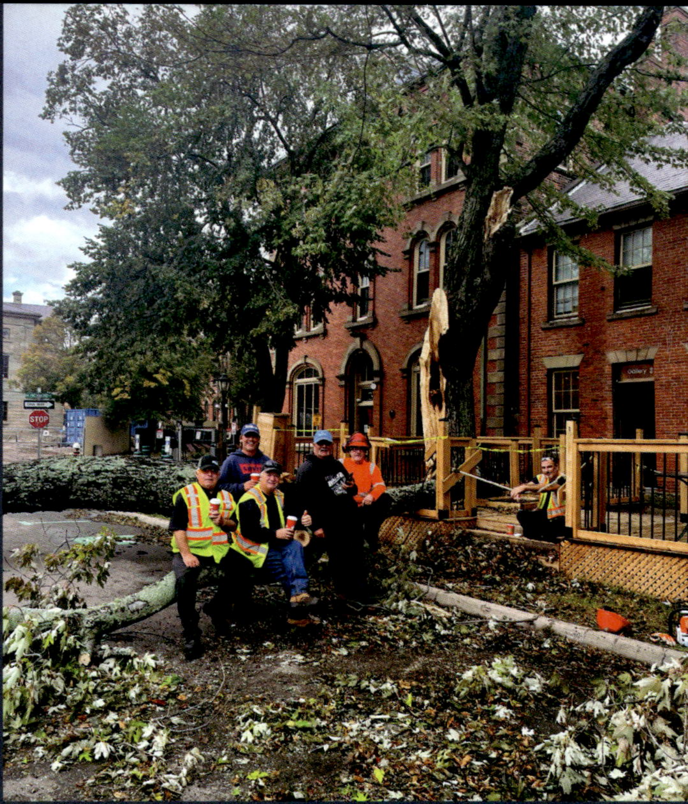

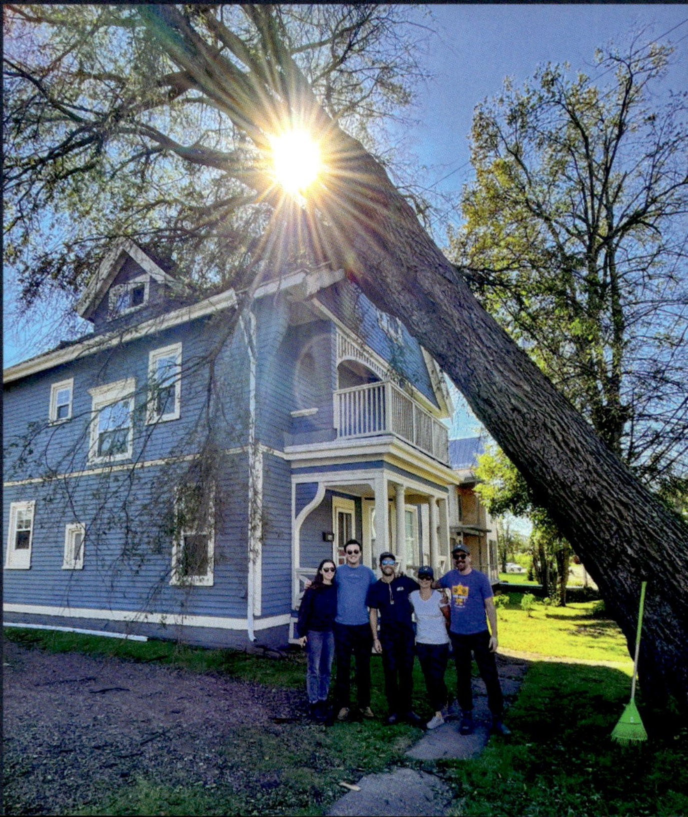

Horses, Cows and Cormorants

FROM THE TINY FRAGMENTS OF LEAVES that clung to the windows to the gigantic swaying trees, Fiona set out to get our attention. Her winds conspired with the rain to tear roofs off, pour water in, soak floors down to the joists, shred walls of gyprock, and crush windows. As glass shattered, we huddled in our homes—some, like me, alone but for our animal companions. As we did so, the close calls to human life were many. But, as someone said, "no one died in the storm," except animals. Fiona dropped barns like a child jumping on cardboard boxes.

Unlike her somewhat gentler brothers, Juan and Dorian, Fiona raged for eighteen hours and took her sweet time to calm down and let the light back in. A few days later on our battery-operated radios, a woman told us the story of that night and the horses in her therapeutic riding stable. She tucked the four of them away on Friday, with as much security as she could provide. But the barn collapsed sometime in the middle of the night. When she found her horses they were not, as she feared, crushed under the weight but stood together close by some trees which, fortunately, did not topple and come up at their roots. These animals, the most beautiful of God's creatures, herded together and survived.

Most of us raised on this Island have seen a cow milked at least once. I think of cows as hardworking and cooperative, despite my childhood experiences of hand-milking a cow who inevitably swatted me across the face with her tail after she lifted it to pee. Cows don't fuss much and they're not known to be particularly impulsive or stubborn. Most farmers on the Island still name their cows, even if they also assign them numbers. To euthanize Daisy and Bessy in the middle of chaotic howling winds would be more harrowing and heartbreaking for a farmer.

This past winter, in 2022, for the first time I can recall, we had virtually no offshore ice from the Northern Gulf and very little shore ice. Made for an early spring, but no winter ice doesn't do much for coastal erosion. This was another symptom of the warming water temperatures that brought the wrath of Fiona down on us. If La Niña's trade winds are kind and bring us ice cover, she will help cool the aftermath of Fiona's meltdown, so the fishermen tell me. They see the impact of warming water through the changing patterns of aquatic species. They don't blame Fiona for demanding we acknowledge her frustration with us humans.

On one of those incredibly beautiful fall days that followed the storm,

I took a walk on the beach below my north shore home. The backside of the dune is still intact and is healthy enough to recover. The winds tore the front, but the marram grass will hold on, as long as we don't get too many heavy winds and acute tidal surges. As I walked along the path, I met a couple with a little girl. She lifted up the shell of a lobster tail so I could admire her find. This thing was huge, too big to get in our traps, but would've been talked about for days if it had wiggled through the hoop: an old grand-pappy of the sea. I wondered what became of the rest of the lobster when the waves dragged it out from under the rocks where it hoped to shelter.

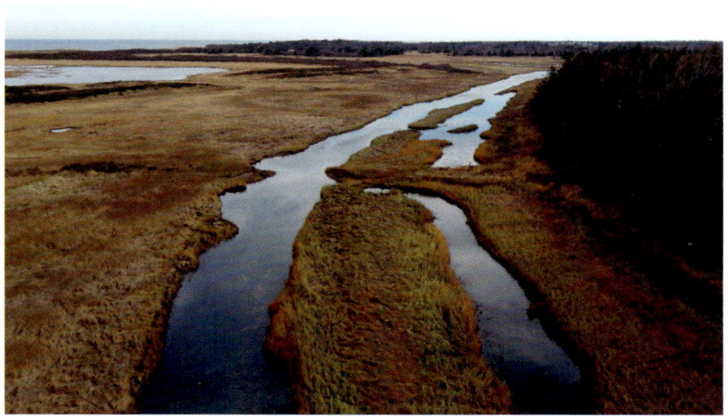

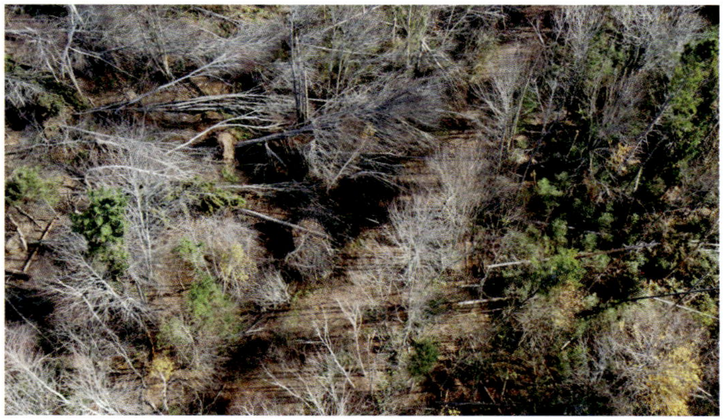

My neighbours are talking. Dead animals on the beach, best to keep the dogs away. At the end of the dune that looks over the channel to Rustico Harbour, pieces of the two houses that swept across from the Harbour are scattered: a roof, part of a wall, scraps of boards tilting against the marram grass, all interspersed with twelve-by-sixteen-inch creosote piles torn away from Court's Wharf. As I approached the wreckage, a black cormorant stood out against the white sand. I had never seen a solitary cormorant at rest, only flying or in a group on wharf pilings, let alone stretched out flat on the ground. Did the wind snap its long neck? The brilliant orange beak against the sand by its lifeless body startled me, but it too will fade in time.

As if in wait of someone to take a seat, a recliner rocker sits upright on the sand dune. Tables too, planted on their spindly legs with a couple of beer bottles nearby that look like they rolled off during a night of casual drinking. Bits of these iconic houses now litter the beach, and much of the frames that knitted them together must be resting on the floor of the bay. We don't need any more damage to deep-keeled boats courtesy of Fiona. A red plastic gas canister by the beer bottles reminds me I will need to get a generator and burn fossil fuels again to heat my house.

When I look across to the capes and over to the sandbars I am, as usual, mesmerized by the softer reds of this land. We have lost a great deal of green where we live, and we miss the fall colours this year. But the blue of the ocean, the red of the soil and the creamy white of the dune sand will still be here. This is our true privilege. We automatically balance it against the fierceness of the wind for a life lived on an island, where we, and our animal cohabitants, are vulnerable to wind and rain and shifting conditions.

Pondering Contemplation and Community Support

THE ENTIRE UNIVERSE AND ITS CREATURES, living and nonliving, are all subject to the command of Allah the Almighty. He wills to whomever He wills, and in that are signs for the believers; but the power of Allah Almighty is strong too.

Although we lost power, we were still very fortunate. The shingles on our roof were damaged and that caused a leak inside the building, as water dropped down to prayer mats. Because of our proximity to the airport, we got power early on. On Saturday of that week, we announced to the community that we would open a warming centre: whoever needed to charge their devices, get some hot water, warm up some food, or even use the washrooms was welcome to our facility. Many people showed up, including families with children, the elderly, and more.

Even though we were limited in our abilities, we were in contact with others and soon organized a generator amidst the disaster. The generator we received will be used as a backup in future power outages, as well. During Fiona, we were able to provide warm food for the community, with some showing up and eating in the basement of the mosque, and some preferring takeout.

We as the Muslim Society of PEI had a responsibility to give back to the community as much as we could. The entire universe and its creatures, living and nonliving, are all subject to the command of Allah the Almighty. He wills to whomever He wills, and in that are signs for the believers; and the power of Allah Almighty is strong.

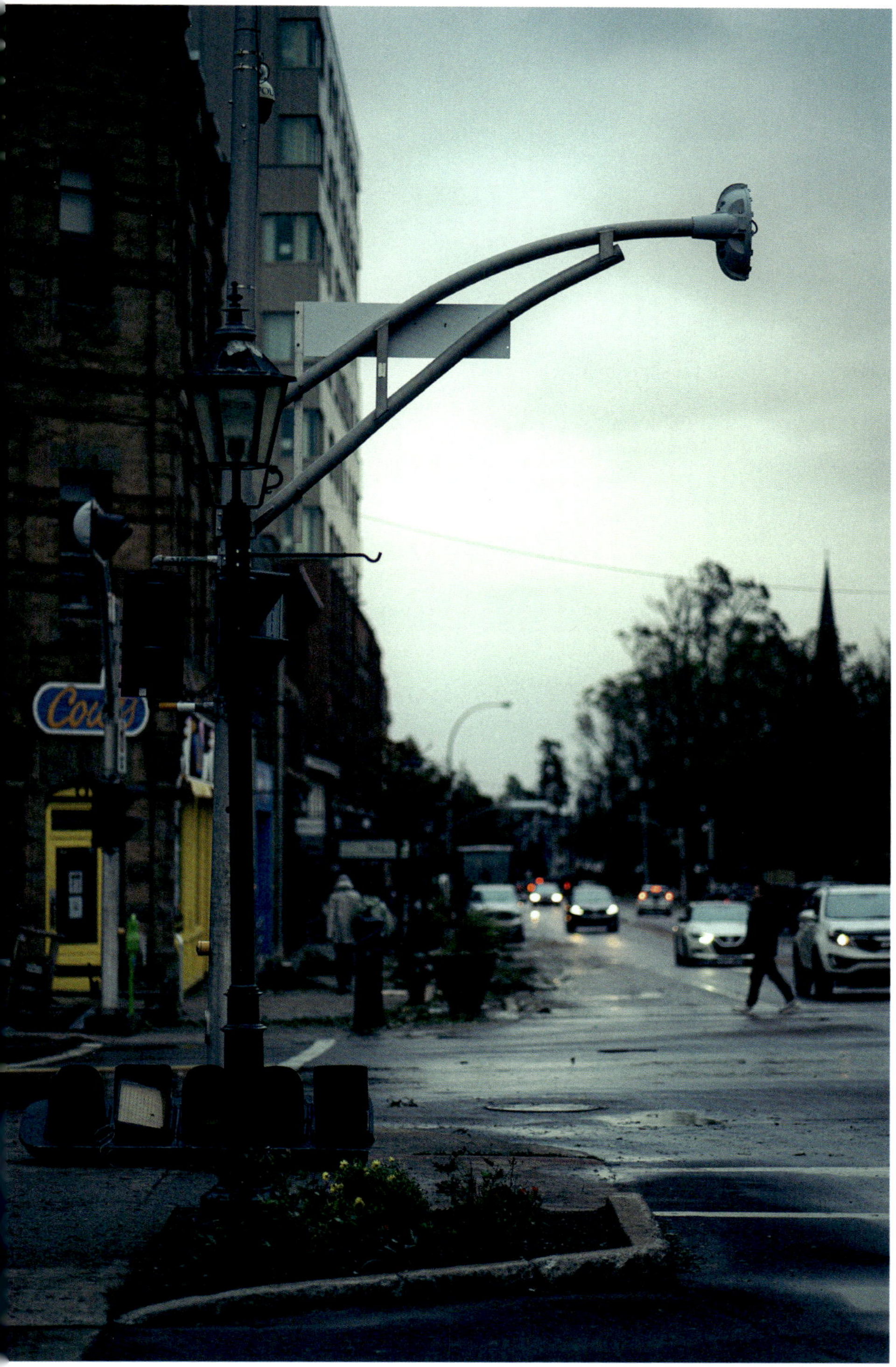

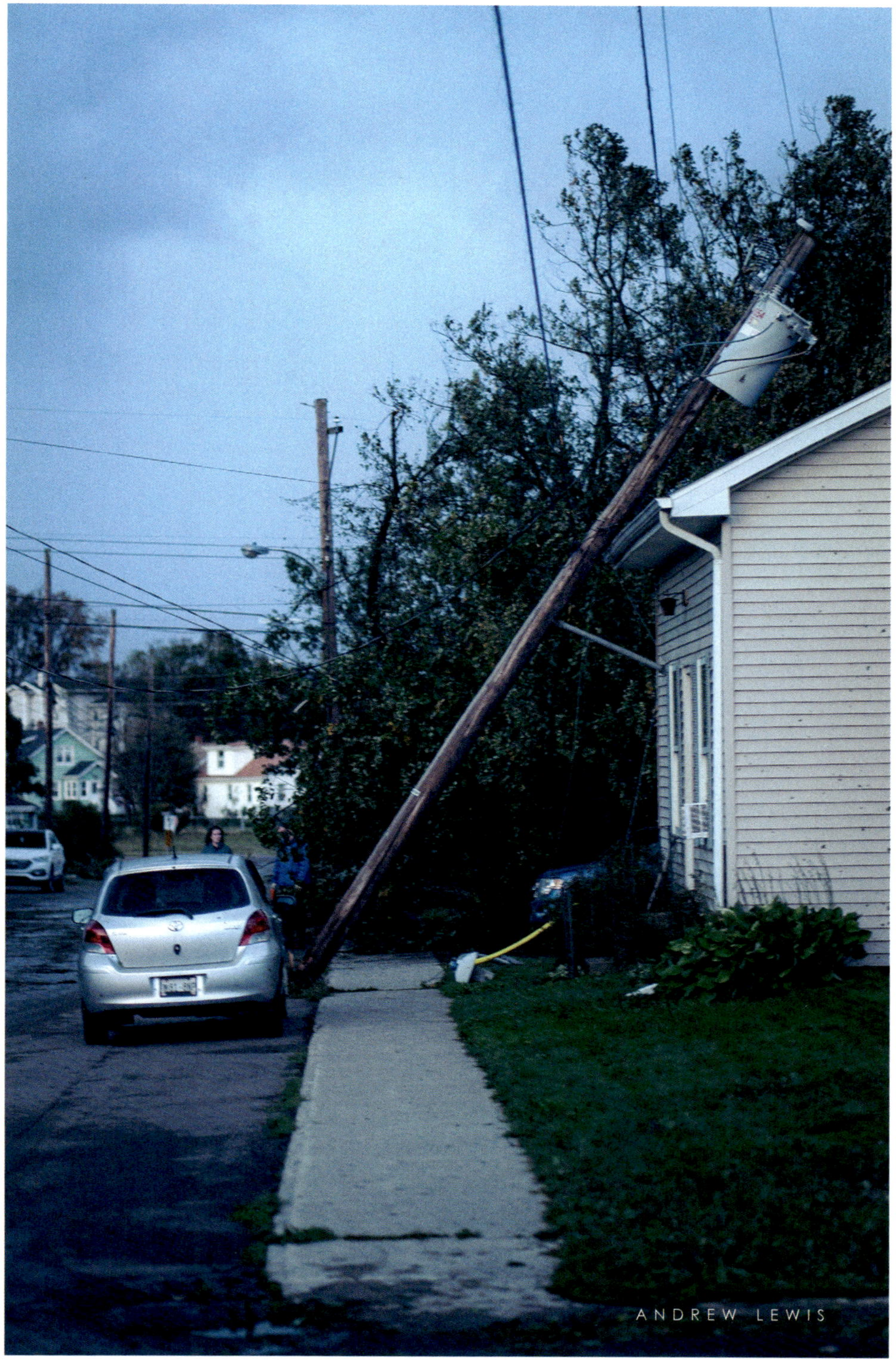

ANDREW LEWIS

A Helping Hand for Those Who Came Before Us

WHEN YOU DRIVE UP MAIN STREET in Montague and head toward Cardigan, you'll pass Sobeys, the Superstore, and, inevitably, the giant red Robin's sign on the right. After you turn, you'll pass houses, pass trees, and pass the Confederation Trail, until you come to a stop at the end of the intersection. Here, you will have arrived at our sacred home. It is our residence: a place where our steps have molded and smoothed the trails we walk on today, where our hands have shaped the landscape, and where we have envisioned the community we are now a part of.

When Hurricane Fiona swept across Prince Edward Island, like Santa Claus, she hoped everyone could feel her presence. Every household was showered with gifts—a confetti of leaves and a lifetime supply of firewood. Our ancestors of Brudenell were not forgotten. And as they could not clean up themselves, we—a future generation forever in gratitude—could not forget them either.

In the days that followed Fiona, our Buddhist nuns drove around Brudenell surveying the aftermath and offering help to our neighbours. Every time we drove by the Brudenell Pioneer Cemetery, we couldn't help but slow down and stare at the obstacle course of fallen trees and branches. It was like a Halloween prank gone awry. But we couldn't just run in and start picking up branches...or could we?

The next day, as we were driving down Brudenell Point Road, we noticed a truck parked next to the cemetery. A few elderly women were walking around the fallen trees. They had an air of purpose about them. We immediately pulled up to the side of the road and ran out to greet them. As luck would have it, we had before us The Guardians of the Brudenell Cemetery.

We learned they were a team of volunteers who helped to maintain the cemetery. There were only a handful of them, all retired. "Were you going to do all of this yourself?" we asked, flabbergasted. Indeed, they were. They are truly angels among us, and strong ones at that. Saws in hand, they were ready to take on the daunting task of cleanup.

We asked if our Buddhist nuns could help. In our tradition, elders, living or past, are honoured and held in great respect. And as we lived just a ten-minute walk down the road, they were our neighbours, after all. They graciously accepted our offer.

And so we found a young man in Brudenell who could cut up the fallen trees. Then, a group of our Buddhist nuns put on our working gloves, grabbed our rakes, and carted in our wheelbarrows. Together, we cleared out the fallen trees and the remains of the storm.

As we were finishing up, we saw one of our Buddhist nuns pacing around a tombstone. There was a fallen tree in front of it. We were going to leave it there as it hadn't been cut up. With our smaller builds, we couldn't possibly move the entire tree...or could we?

"It's a bit oppressive, isn't it?" The nun said. "Having a tree blocking the front of your doorstep?"

We paused. Indeed, that would be quite uncomfortable. If it was our home, we would want more space. We called all the nuns over, and a dozen or so of us stood around the tree. People shouted orders of coordination, from left and from right. A voice rang out as it gave commands. Perhaps it was the sheer power of mind, or perhaps it was the extra hours of training from our recent tug-of-war competition—but that tree sure did come off the ground!

In the days that followed Fiona, we witnessed how the Island defines "community." Neighbours checked in on one another, restaurants offered free hot food, businesses sent employees out to help, people opened their homes to strangers, offering warm showers and pulling extension cords out onto their lawns as makeshift charging stations. Island hospitality is defined by a timeless tradition of warmth carried on by generations before us, and that we hope to continue for generations to come. We are grateful to those who came before us, for the beautiful Island they left us that we now call home.

GREAT WISDOM BUDDHIST INSTITUTE

Lemons to Lemonade

FOR THE PAST TEN YEARS, our Buddhist nuns have called Prince Edward Island home. We've experienced a bit of PEI's uniqueness and how her quiet charm somehow draws people together. The Island never fails to bring out the best in us; even if it is something small, you can always count on someone to lend a helping hand. Day or night, our neighbour across the road answered our calls with his usual jovial personality, coming to our rescue with all sorts of gear—including his tractor.

As you can imagine, most of our Buddhist nuns come from warmer parts of the world. Could we survive the rugged Arctic winds and chilly winters? How does one keep warm, drive, or even plough snow in the winter? And what about that Island twang? How do you walk on those "slippy" roads when it's "something cold" outside? We found out the Island way: Like everyone else, we go to our neighbours. Tap into their inexhaustible silo of life experience for the best answers.

In the past ten years, the friends we've made have become like family, just like the MacLeod farm's hundred-year-old linden trees. They are very much an integral part of the landscape. When our Buddhist nuns see the trees in the driveway, we know we're home. They are also the home to a lively community of sparrows, blue jays, woodpeckers, and squirrels. They have a quiet and stately grace, sway elegantly in the gentle winds, and wave friendly hellos.

Over the years, many guests to the monastery have stopped upon arrival to admire these towering giants that have spanned generations. As for curious little ones, they nimbly explore the trunks like aspiring Olympians, swinging by the branches of Nature's jungle gym. With notable Halloween guests like Elsa from *Frozen*, Minions, witches, monsters, goblins and other honourable trick-or-treaters from Vernon River Consolidated School, many have paid special tribute to these trees; the monks visit as well, of course, often with candy in hand.

On days when there's a cacophony of sounds clamouring inside, there's nothing like standing by these linden trees. They have always brought the serenity of harvest nights, and a quiet moment for those who look to stay out of the limelight. On blistering summer days, we have found respite in their cooling shadow. Oh, linden tree! Through all of our highs and lows, the

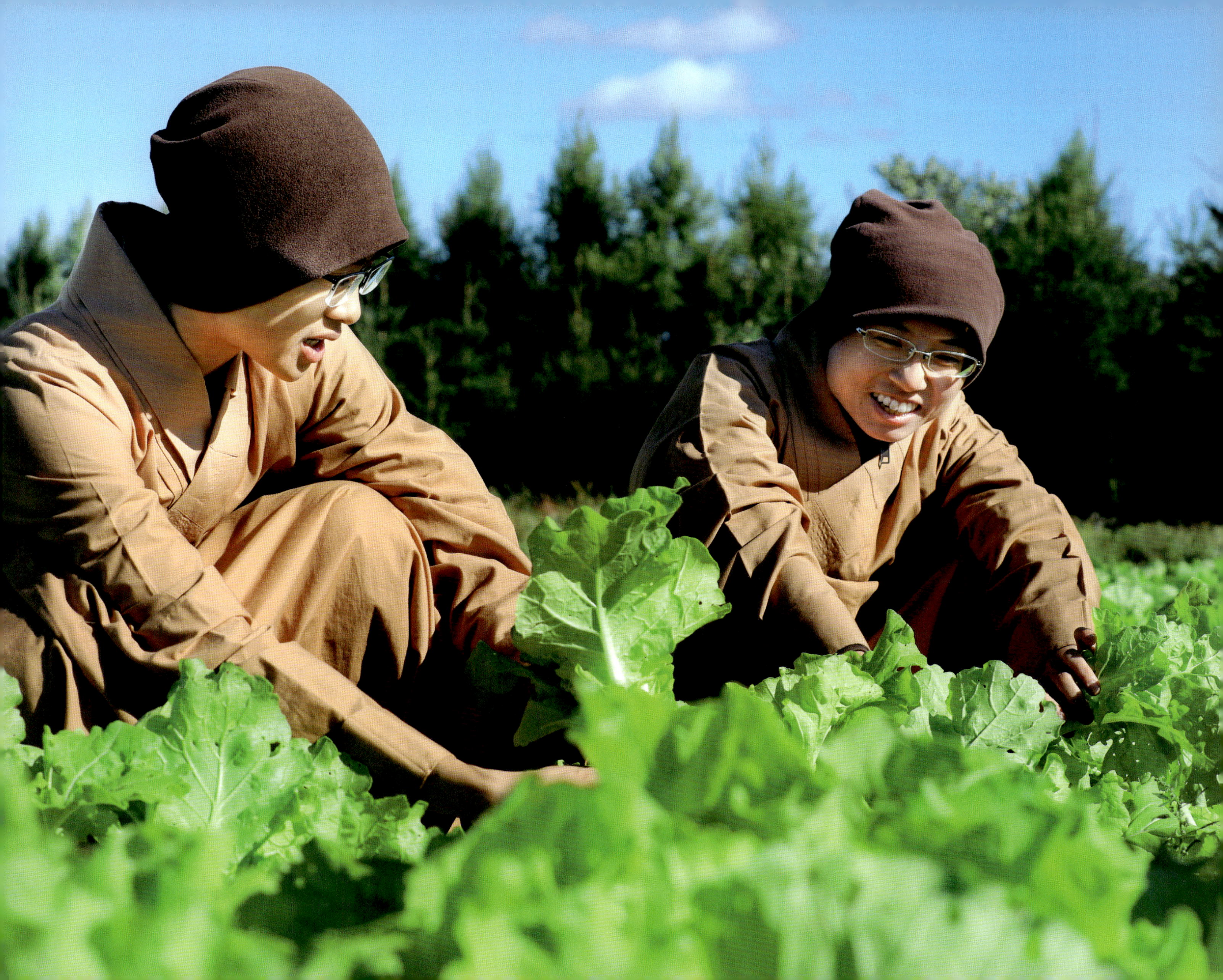

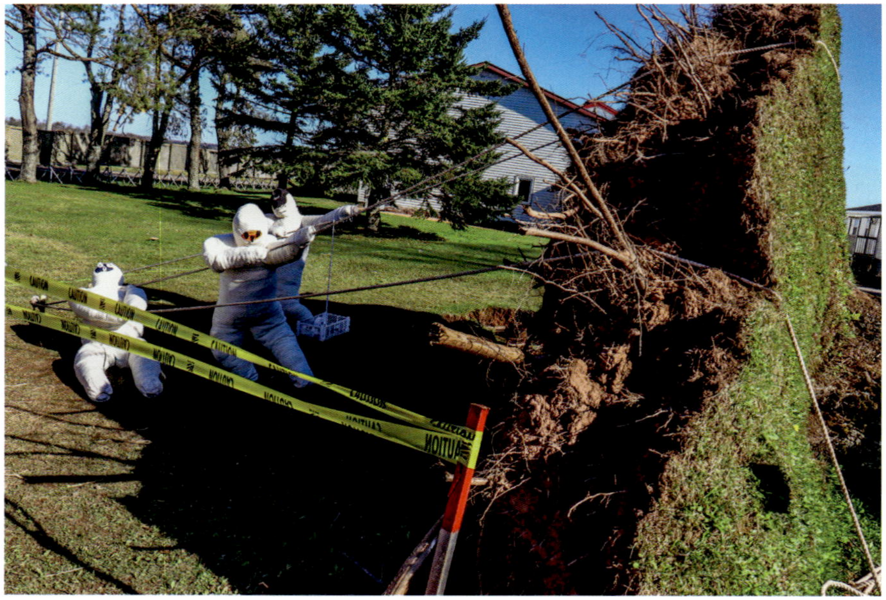

ebbs and flows in life, you've been with us through them all: a true friend.

On September 24th, 2022, Fiona marked her presence with blaring winds and cutting machetes, ruthlessly slicing and dicing casualties everywhere. Her winds left one of our hundred-year-old linden trees as a lifeless, barren log on the ground. The tree was no longer home to hundreds of creatures, no longer tall and steadfast, and she had lost her place as a pillar welcoming all of our guests. Each Buddhist nun walked past this linden tree with a prayer and heavy heart. Our resilient friend fared the toughest stretches of history, both world wars and Covid, but alas, its time had come.

Besides heartache, the loss of an old friend, a multitude of firewood, and a newly fenced-off four-foot-deep crater, what else was there? As our Buddhist teacher says, "Even a wilted flower has its worth, providing for birds in winters' dearth." Creativity is an art which can be expressed through thoughts, language, and craftsmanship. Besides focusing on loss, what lateral thinking could we engage in? Like the Island Waste Management Corporation, could we change how we see Fiona, from that of a catastrophe into something reusable, recyclable? How?

A light bulb went off: Halloween! What better backdrop to use than the aftermath of Fiona, and what better opportunity to send messages of hope, warmth, and community? Filled with anticipation, for the next two and a half weeks, after clearing out the storm debris, our Buddhist nuns planned, designed, and decorated a spooktacular scene for all to visit on Halloween: a danger zone with zombies desperately attempting to pull up the fallen linden tree, adjacent to two creepy jack-o'-lantern-men with changeable, glowing pumpkin heads, a massive magical broom made from fallen twigs and branches, and a ten-foot-tall alien from *Toy Story* welcoming trick-or-treaters. After throwing in some handmade cut-outs, stitched-together pieces of cloth, bales of hay and glowing lights, it was a trick-or-treater's paradise!

Laughter, smiles, amusement, and joy abounded that evening, all of which came despite the backdrop of chaos. Over a hundred trick-or-treaters visited us: both young and old, families and friends, and of course, our wonderful neighbours from across the road. Fiona may have swept through the Island on September 24th, but she couldn't quell the Island spirit, our resilience and togetherness. Though we lost many trees and power for days on end, Fiona brought everyone together. At the end of the day, those connections and gifts are what matter the most.

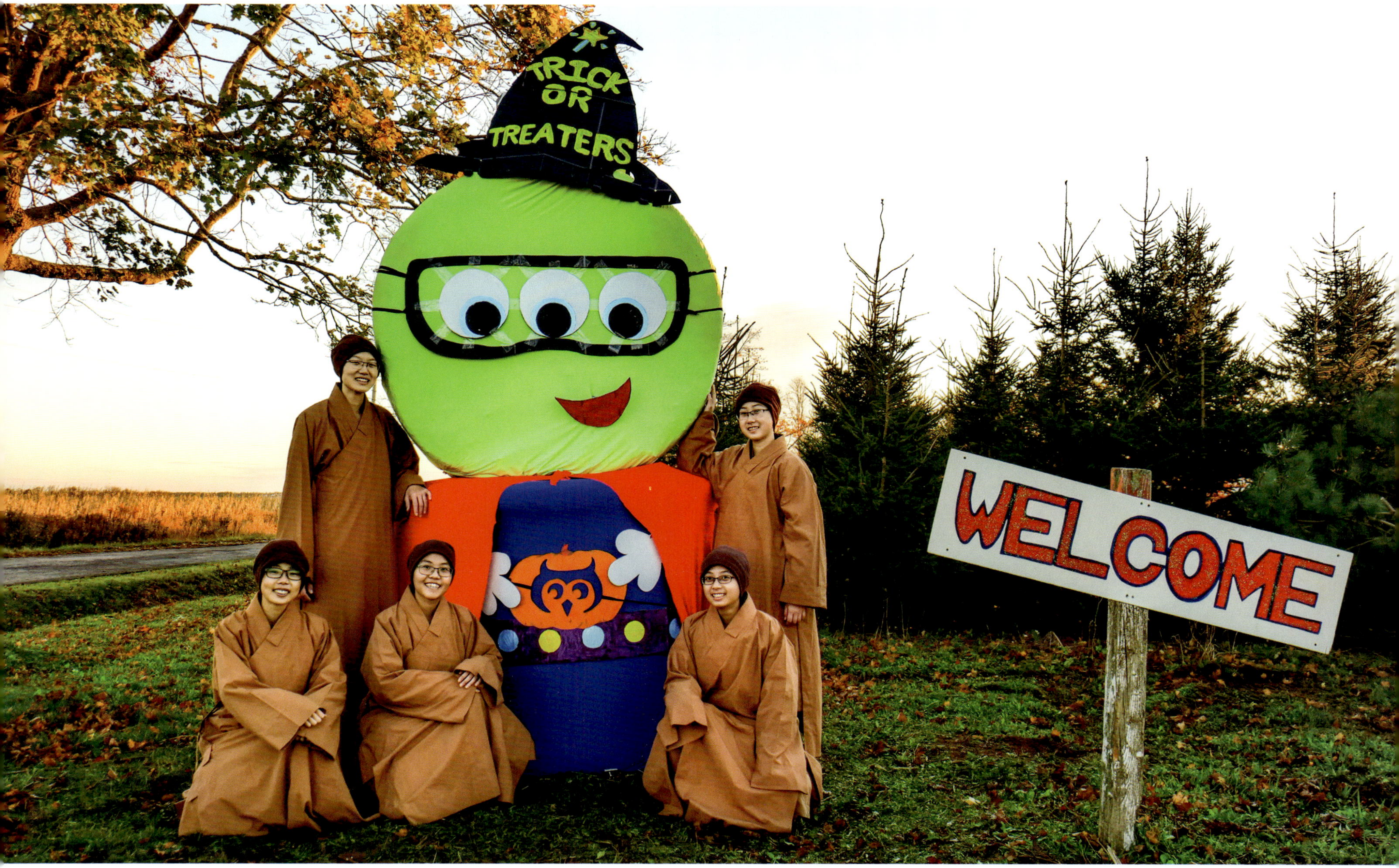
TRICK OR TREATERS
WELCOME

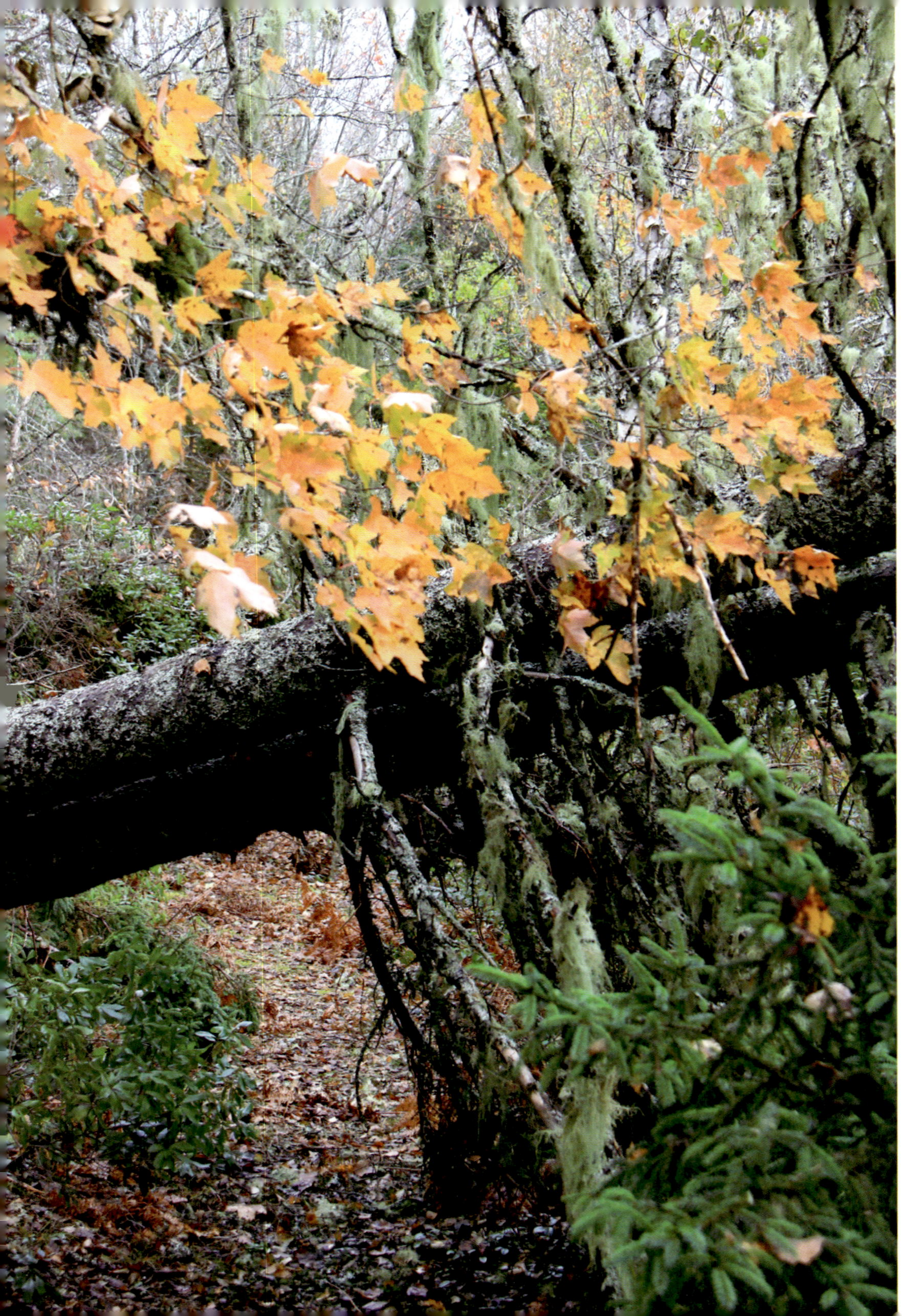

DIANE HICKS MORROW

Like Two Slant Trees

"Lean on me," he said,
loving her weakness
and she leaned hard
adoring his strength.
Like two slant trees
they grew together
their roots the wrong way
for standing alone.
—Fred Cogswell

WE CHOSE THIS POEM BY MY FAVOURITE PROFESSOR for our marriage ceremony, fifty years ago. Four years later we settled into this old farmhouse in West Covehead, on twenty-five acres of half-wooded land, overlooking a stream and pond. At first glance, the twin lindens on our front lawn appeared to be one huge tree instead of two, their trunks only a hammock-length apart. Not identical twins: one trunk was taller, thicker. They were a pair, like us.

Severe ice storm damage to the smaller tree in a nor'easter twenty years later only increased their difference. I stood in the living room bay window, helpless, as crystal-coated branches clacked and creaked in the wind, then crashed to the glassy snow below. The next day, I let tears flow while I pulled monstrous fallen branches along the ice to the edge of the lawn, sliding the weight of the branches that neither of us could lift, and before the sun could melt it.

Over the years, the smaller tree slowly filled in, except for a welcome keyhole allowing me to peek at Uranus from my side of our bed. By midsummer this year, my view of that "rebel angel" planet had become blocked by the lushest linden leaves ever. Opening my eyes each morning now, I feel aloft in a floating treehouse.

From the hammock, as I look up into their canopies, I see how fairly these century lindens share available light and space. After fifty years, my husband and I have learned to do the same. We reminisce about past gatherings in linden shade: celebrating our elders' lives with family and friends; taking our oldest son's wedding photos beneath umbrellas in the rain; reading to grandkids in the hammock. We talk about celebrating our fiftieth anniversary (and our lindens' 100th birthday!) with an outdoor gathering beneath their parasol of shade, but we've not yet chosen a date.

Bracing for Hurricane Fiona ends any celebration talk. We remind ourselves how little damage hurricanes Juan and Dorian caused to these stalwart sentinels. Now, with wind gusts approaching 200 km/h and the power down by midnight, we can see nothing outside. From our bed, with the roar of an invisible jet squadron circling our quaking farmhouse, we can also hear ominous thuds we've never heard before.

At dawn, through kitchen windows freshly "pestoed" by mashed linden leaves and opaque slime, we can barely make out some raw wood on the front lawn. Could it be plywood torpedoed from our neighbour's half-built shed? The howling gale prevents any foolhardy attempt to open a door and walk across the yard to investigate. But as daylight brightens, we finally see from the living room bay window—the long raw gashes along the upper main branch of the thinner linden. Its severed crown is splayed across our front lawn. It has narrowly missed all front windows, but blocks access to the oil tank fill pipe.

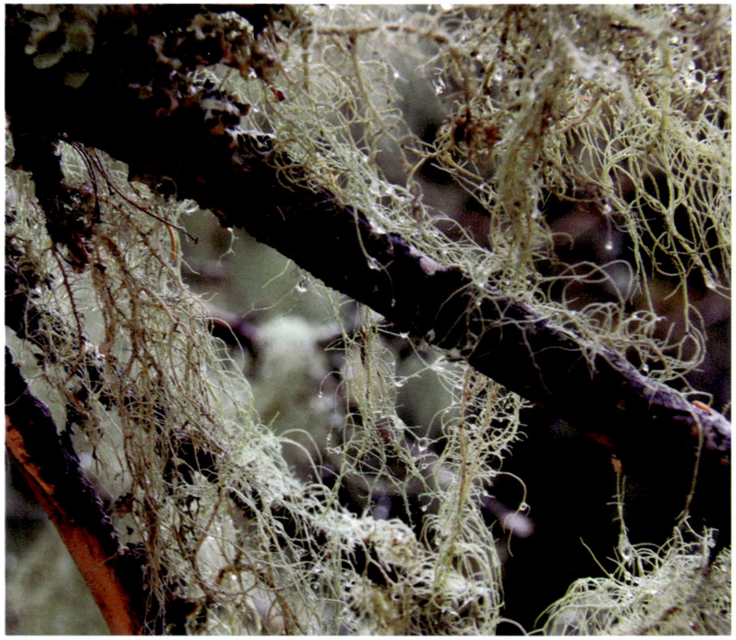

The frightening gusts persist until nearly dusk on that long, long day. More tree branches are mangled and falling, while many are left hanging in tortured angles high in the trees. Clearly, Fiona has wounded the lindens more than any previous storm.

At first, I feel gratitude that both are still standing. By dusk, my first walk into the yard numbs me beyond tears, until the aroma of the sugary sap flowing from the linden's assaulted trunk overwhelms me. But then, I observe that the roots have held.

We lose power for ten days. We shift into survival mode in our house, lacking water and central heat. The unseasonably warm weather means we rely on generous neighbours with generators to save our frozen food and refreeze our fridge ice packs, as well as to energize the charging brick for our devices. What a relief to text with concerned friends across North America and Mexico. Many of them, ironically, know more about what's happening across Atlantic Canada than we do; news of the great Fiona has spread. The wind-up radio is our lifeline to sporadic local information.

Now, over a month later, the full extent of our tree loss overwhelms us. It feels like I'll never be able to navigate the debris-buried slope down to the rowboat and dock, assuming they are even still intact. A heartbreaking number of the trees surrounding our lawn are fallen or staggering like drunks.

The family woodlot on the Friston

PHOTOS BY THE ISLAND NATURE TRUST, JULAIN MOLNAR, SARA DEVEAU, DAWN MACDONALD

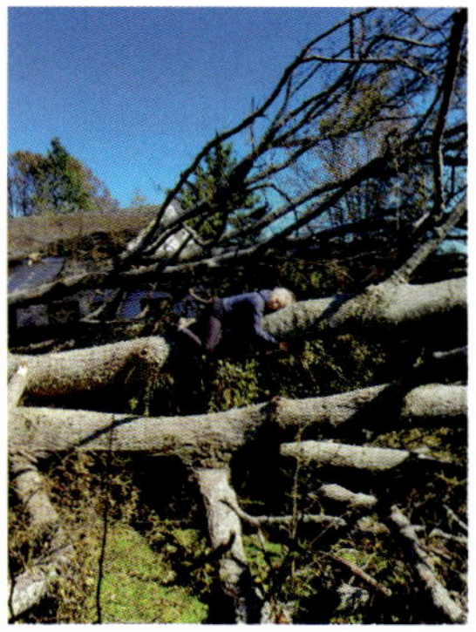

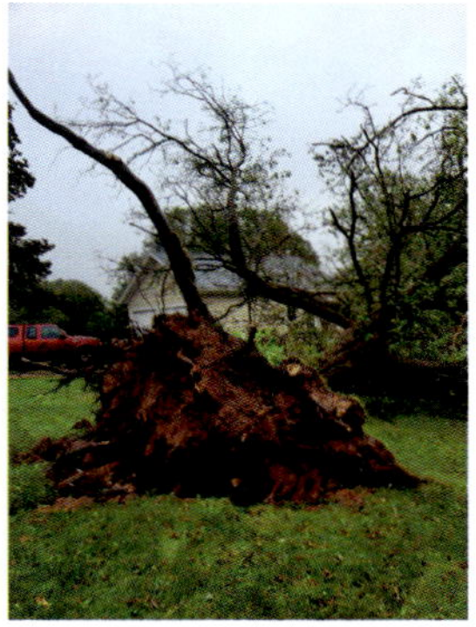

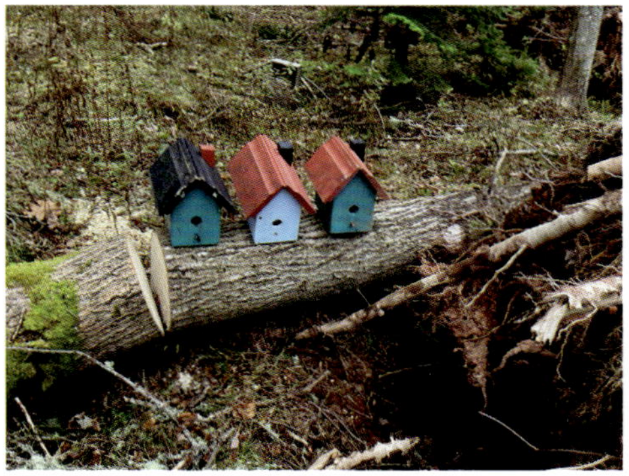

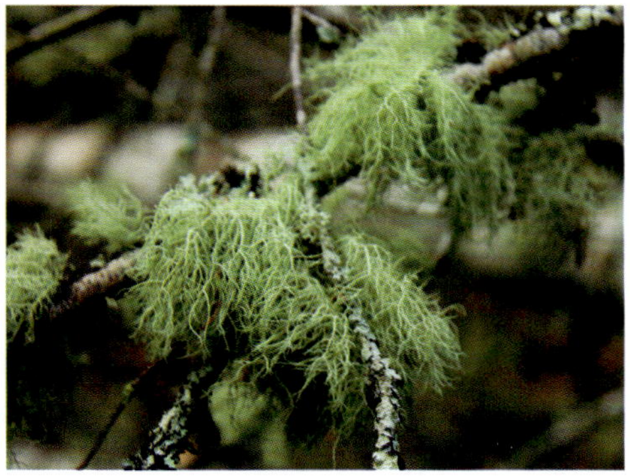

Road looks like a war zone. Our younger son's house stands in the middle of a fallen pine plantation there. He's been chainsawing daily, with no end to this exhausting labour in sight.

And yet, and yet. Is it possible that his pines and our lindens have borne the brunt of the storm and saved our homes from serious damage? This week we've been surprised by how many trees are still standing tall. And we're comforted to once again spy again elusive Uranus from our bed—and even more stars we could not see before.

Now, as Andy and I are learning to lean on each other for balance, I revisit our wedding day poem. When we married, I'd been living with a weakened polio leg and foot all my life. I soon became gladly wedded to Andy's strength. But about twenty years ago, when his neuropathy began to show, he could no longer get into the rowboat. Later still, he could no longer manage the steep path to the pond. Now, as he uses a walker outside the house, and a cane inside, I am becoming the go-for and do-er for both of us, grateful for my one strong leg.

Indeed, the poem's final two stanzas sum up the reality of our bodies:

Like two slant trees
they grew together
their roots the wrong way
for standing alone

We are battered but not beaten, as our revised version shows:

Like two *lindens*
they grew together
their roots the *right* way
for standing *paired.*

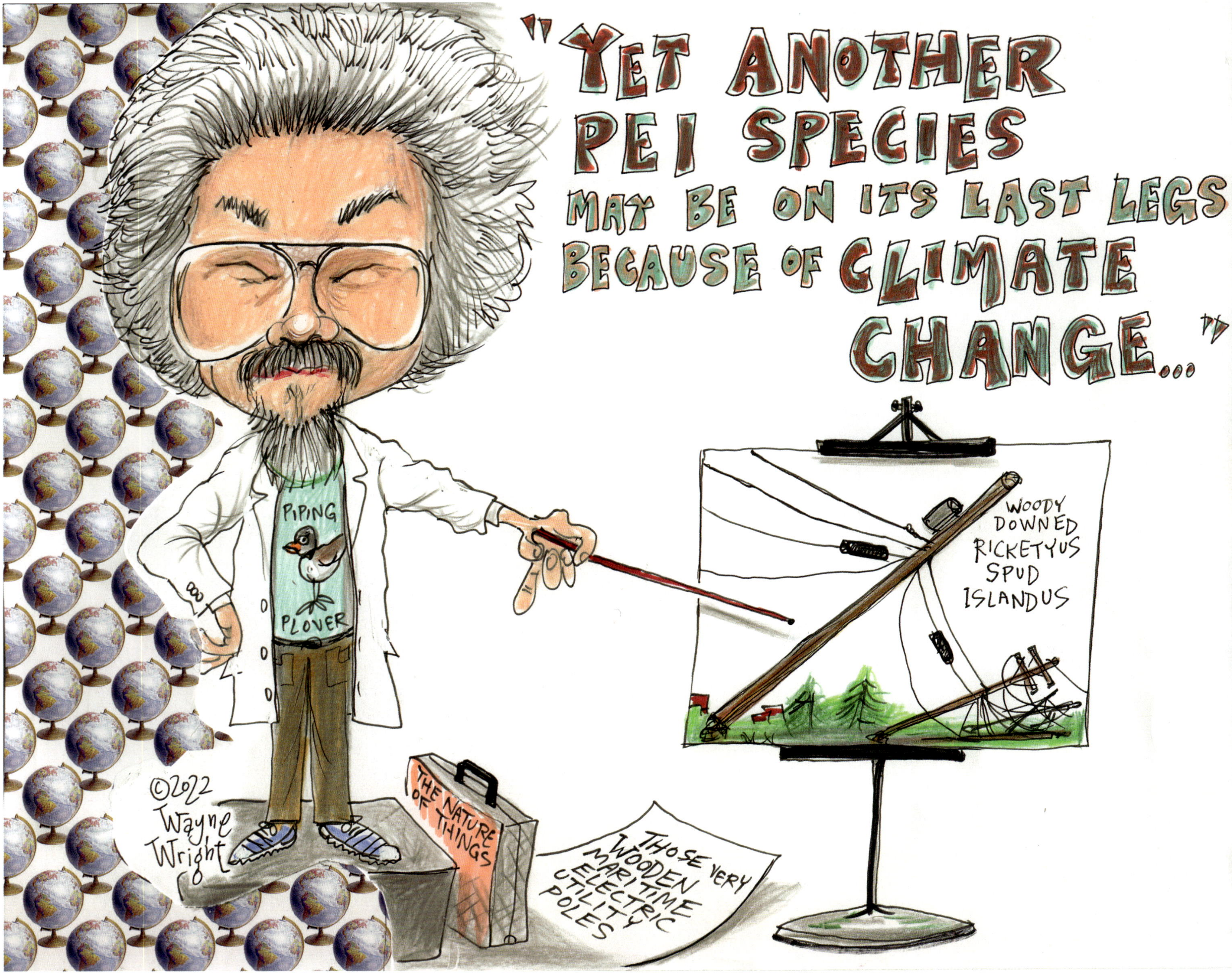
"YET ANOTHER PEI SPECIES MAY BE ON ITS LAST LEGS BECAUSE OF CLIMATE CHANGE..."
PIPING PLOVER
WOODY DOWNED RICKETYUS SPUD ISLANDUS
THE NATURE OF THINGS
THOSE VERY WOODEN MARITIME ELECTRIC UTILITY POLES
©2022 Wayne Wright

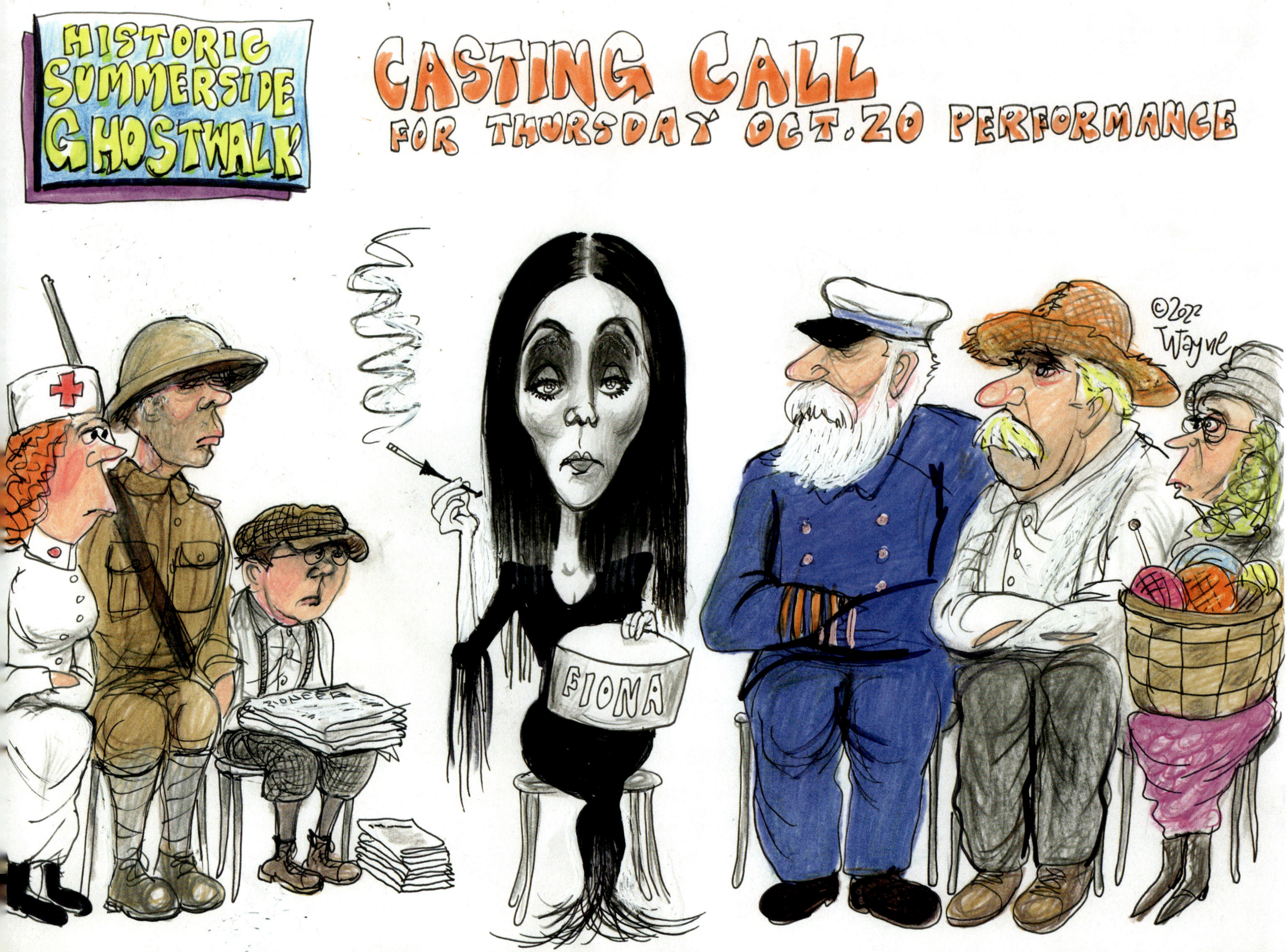
HISTORIC SUMMERSIDE GHOSTWALK
CASTING CALL
FOR THURSDAY OCT. 20 PERFORMANCE
©2022 Wayne
PIONEER
FIONA

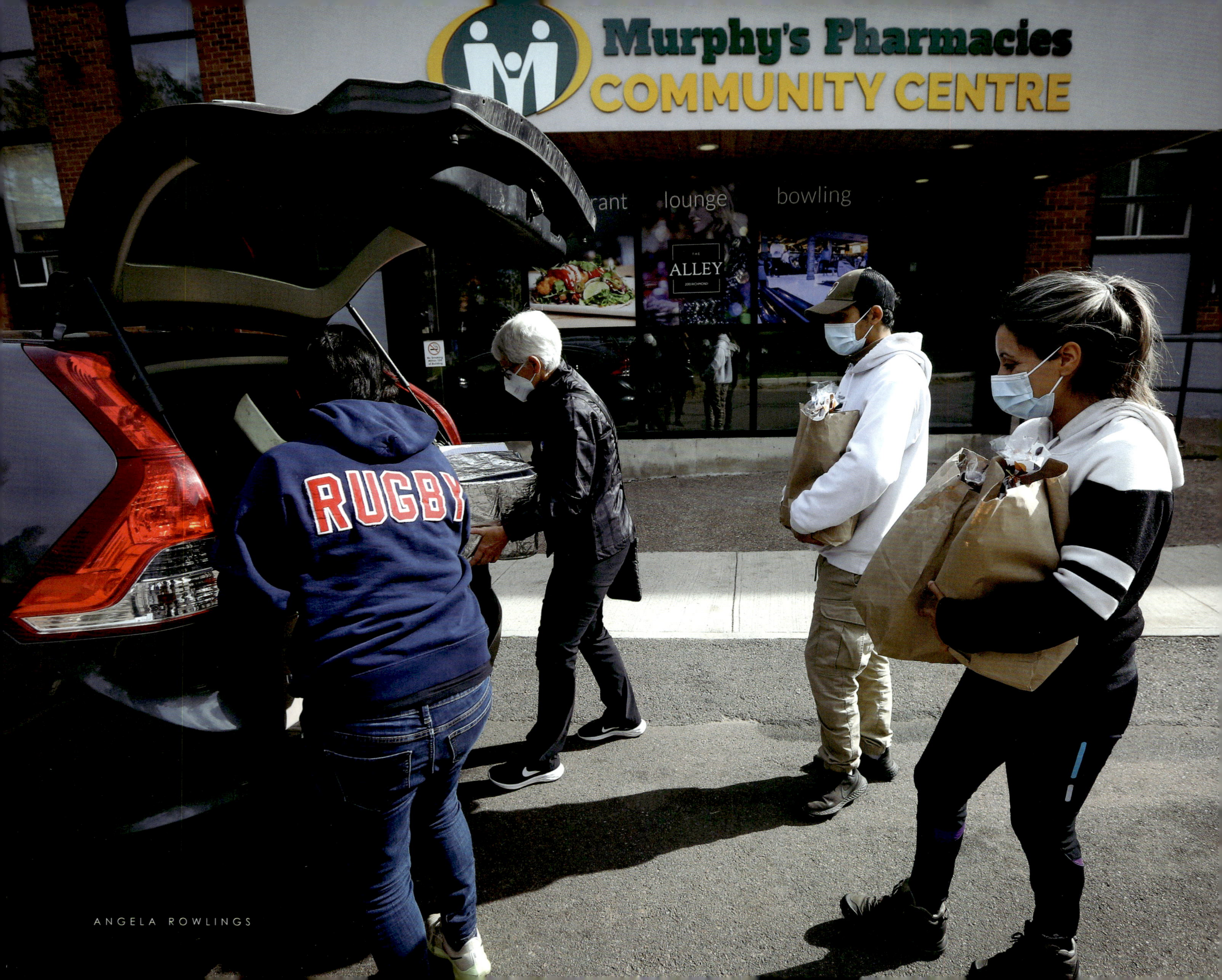

ANGELA ROWLINGS

COVERS PRINCE EDWARD ISLAND LIKE THE DEW FRIDAY, OCTOBER 14, 2022

saltwire.com

The Guardian

CHARLOTTETOWN, PRINCE EDWARD ISLAND A member of the SaltWire Network

CLIMATE CHANGE

Early signs of success

Eco-friendly shoreline protection experiment appears to have worked after Fiona

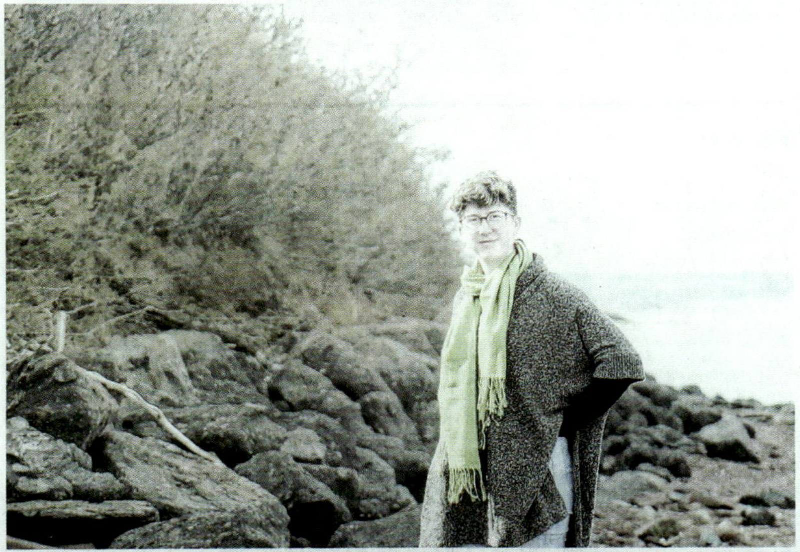

Charlotte Large, project manager for the Stratford Area Watershed Improvement Group, stands by the living shoreline the group built last year at the Stratford Harbourfront on Oct. 13. **RAFE WRIGHT • THE GUARDIAN**

RAFE WRIGHT
LOCAL JOURNALISM INITIATIVE
Rafe.Wright@saltwire.com
@wright542

STRATFORD, P.E.I. — An eco-friendly project with the goal of protecting P.E.I.'s shorelines from erosion in Stratford seems to have survived one its first big tests with minimal damage.

Charlotte Large, project manager for the Stratford Area Watershed Improvment Group (SAWIG), visited the site one week after post-tropical storm Fiona hit the province to assess the damage of the living shorelines the group spent two years building up around Stratford's vulnerable coastal areas.

The living shoreline on the Stratford harbourfront saw the most damage, but much of the infrastructure survived, Large told SaltWire Newtork during an interivew on Oc.t 13.

"What we've seen post-storm is a lot of scouring and erosion at the lower end of

SEE EARLY ON A2

ADRIAN MCNALLY SMITH

Aftermath of Fiona

IN THE AFTERMATH OF FIONA, every Islander has their stories—for the destruction and devastation was so prevalent. Fortunately, there are many stories containing more than just despair. Some of these stories are of epic proportions—harrowing tales, like my friend's property that was destroyed by fallen trees, and her neighbours who showed up unannounced with tractors and chainsaws. Clearing her lane took days, not hours. Or the friends and family members that formed a human chain in Cymbria to rescue some elderly people whose cottage got washed away in the storm. And those stories, thank God, are everywhere.

But sometimes the more subtle messages, though much simpler in scale, are equally meaningful and worthy.

During the COVID pandemic, when my friend's husband was dying, he spent hours in his remaining months watching the birds at the feeder that hung from the tree outside his makeshift bed in the living room. That Christmas, my friend put solar lights on that tree, which had stopped working many, many months ago. She merely left them there, adding electric lights around them. During Fiona, like many Islanders, my friend, now a widow, woke to a morning of darkness, except for the solar lights that had come back on in her husband's tree. She wrote to me saying, "After this happening, one has to believe in something".

The next day the storm raged on. My neighbour's horse, Jack, was ill, choking—and a vet had to be called in. We were all in the barn assisting the vet as best as we could, without the aid of power. At one point, the vet and his wife had to leave to collect their things. My neighbour was exhausted. He leaned his weight back against the stable, and then Jack leaned his large head gently against his friend. At that moment, nothing else mattered than the love between the two: a man and his horse.

These days, kindness has been a way of life. Maybe with the silence between us, the community's good news is able to travel farther. One young mother wanted to help out where she could, though she had a two-year-old and a baby at home. She heard the local orchard had been hit hard, so she gathered her children and headed over. The oldest picked apples for a few hours, and afterwards, thanked her mom for letting her help. Intuitively, at two, this child knew that this was where she was

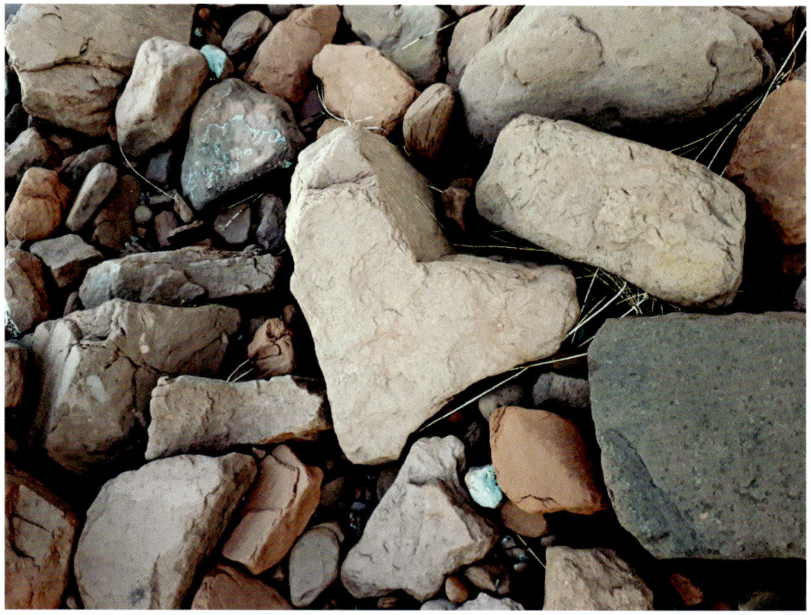

meant to be. We give because we want to and the giving is going around.

I pulled a back muscle cutting down trees. The next day, while we were rounding up the neighbour's calves who had broken through the downed electric fence, another neighbour noticed my discomfort. "Pop over to the house; I have some muscle relaxants," she said. Work schedules don't stand in the way of visits when we wake in the light and sleep by the moon.

After days of helping out and no hot water for showers, I was sore and tired and dirty. In the middle of this madness, I noticed my neighbour's quiet art of photography, as a picture of her daughter as a child graced the wall. The beauty took my breath away—and it reminded me of the power, the gift, and the necessity of art in a time of darkness and despair.

Then I lost my breath for a second time. Though my friend tried and tried to prepare me, I wasn't ready to see North Rustico Harbour. One of my favourite places, the landscape has now been totally changed. As I walked around with a heavy heart, tears collected in my eyes; nothing felt the same. Even the shoreline from the breakwater to the pond had been altered. The dunes had given way, and midway a mass of rocks and stones now divide the two ends. My friend and I walked the shore anyway, and midpoint she spied a huge heart-shaped rock. That rock woke something dormant within me. I learned that I can't love this area any less—it's the spirit of the place, the energy that I adore.

As we left the shore and headed into the village, families of geese and ducks swam in the neighbouring pond; in spite of the surrounding devastation, there was a prevailing sense of peacefulness about them—present and mindful—their calls carrying through the silence of the night, retelling me how to live my life. Sometimes we are just being asked to listen.

JULIE PELLISSIER-LUSH

The Trees

I was always taught to hug the trees.
Watching them change colours, our Island so beautiful.
Seeking the ones that smell familiar, like Christmas in every branch.
My heart aches for our trees.

A storm that changed our island,
an experience we did not need.
Places I knew before are different, empty, broken.
From eager saplings to the great oaks,
knocked down to the ground.
My heart aches for our trees.

Do you know they talk to one another through their roots?
Laying on the ground, you could almost hear their screams.
For a lifetime they grew
protected on our island, growing up from the red clay.
My heart aches for our trees.

And now I mourn, as we see what's left behind.
Roots above ground and old trees that now lean.
Landscape changed, feeling like a mother's empty nest.
Whispers from those roots still haunt my dreams.
My heart aches for our trees.

JULIE BULL

spir(itu)al rEVOLution

science says: we must see it to believe it
spirit whispers: we must believe it to see it
which will you pick?
straddling multiple worlds
albeit imperfectly
capable of realizing every dream we've ever dreamt
but we paddle against currents that we have created
we segregated ourselves from spirit
this ego haze, we need to clear it
the magic you feel
when you look at the art in this book
that's you
your spirit reflecting itself in your being
you're seeing yourself differently now
connected, integrated
as our landscapes are desecrated
so are we
grappling with things our ego cannot comprehend
battling our spirit's need to transcend
we bend til we break/down
like those trees we lost to fiona
we have collectively built this persona where we forget that we are the land
the pain, the gasps, the laughs, the tears, the tingles
reminders that you are alive
you are here
this magnificent being
seeing yourself
perhaps for the first time
this is significant
sacred medicine
be like salmon
return to the place of your creation
it's spacious, there's restitution
every day a chance to be brand new
because the revolution is you

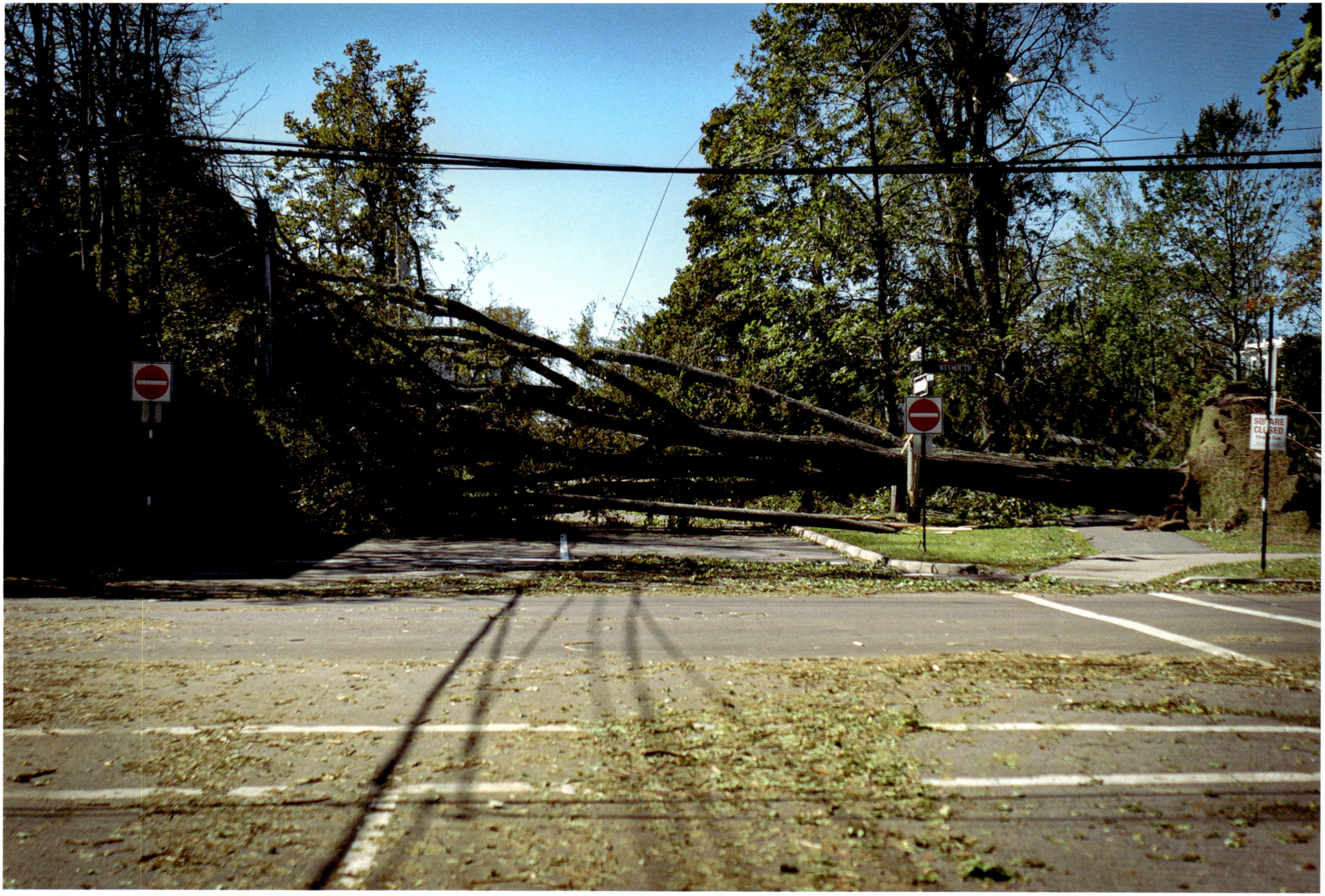
WEYMOUTH
CLOSED

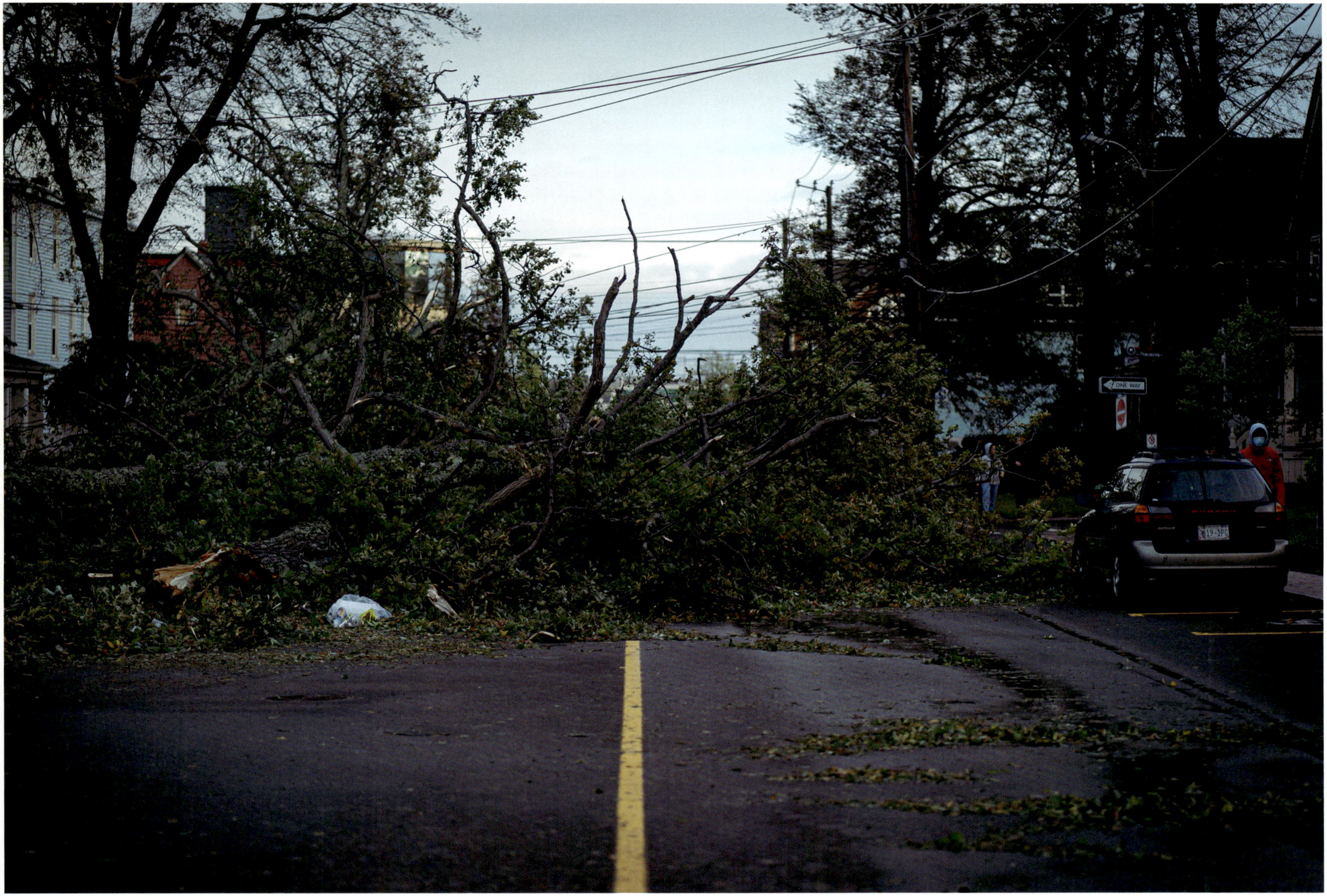

ANDREW LEWIS

OCT. 16 — Wondering why the leaves are so late to turn colour this year. Then it dawns on me: They're not slow to turn, it's just that most of them have blown away.

OCT. 20 — The art of woodpile stacking! Sad testaments but also beautiful reminders that order can come from chaos.

OCT. 25 — I realize I haven't been back to sing to the seals yet. So much to tell them.

OCT. 27 — I walk in my forest more than ever, daily on my twenty-five-step path, and bit by bit I stumble farther in, through heartbreaking debris, constantly surprised by how much more I see and hear. How many deciduous trees have been freed, now that hundreds of spruce are horizontal? The saplings I never noticed before are actually smiling (I swear) at the gift of all that sunlight. We can help each other survive.

OCT. 27 — I am more in love with this island than ever.

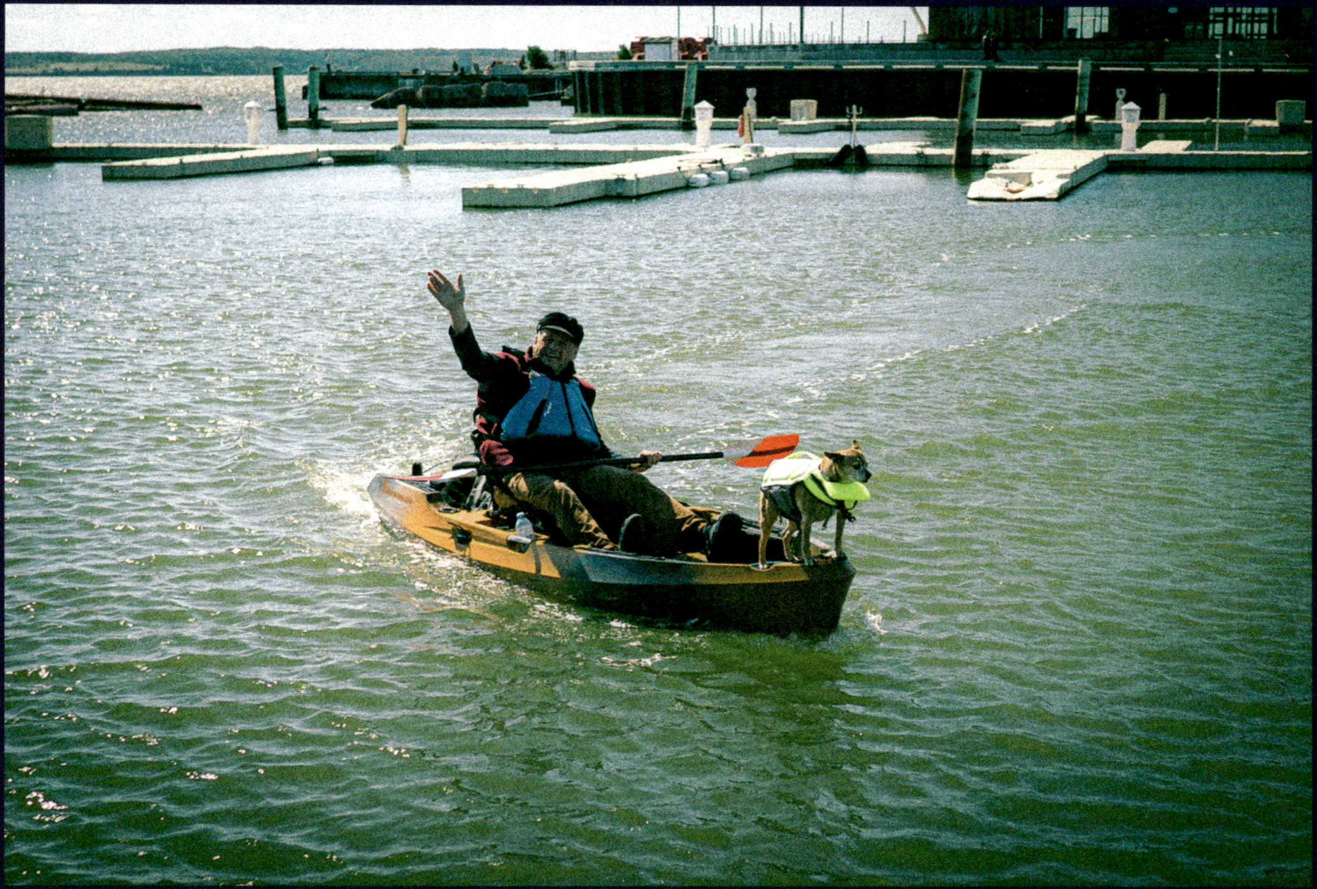

ANDREW LEWIS

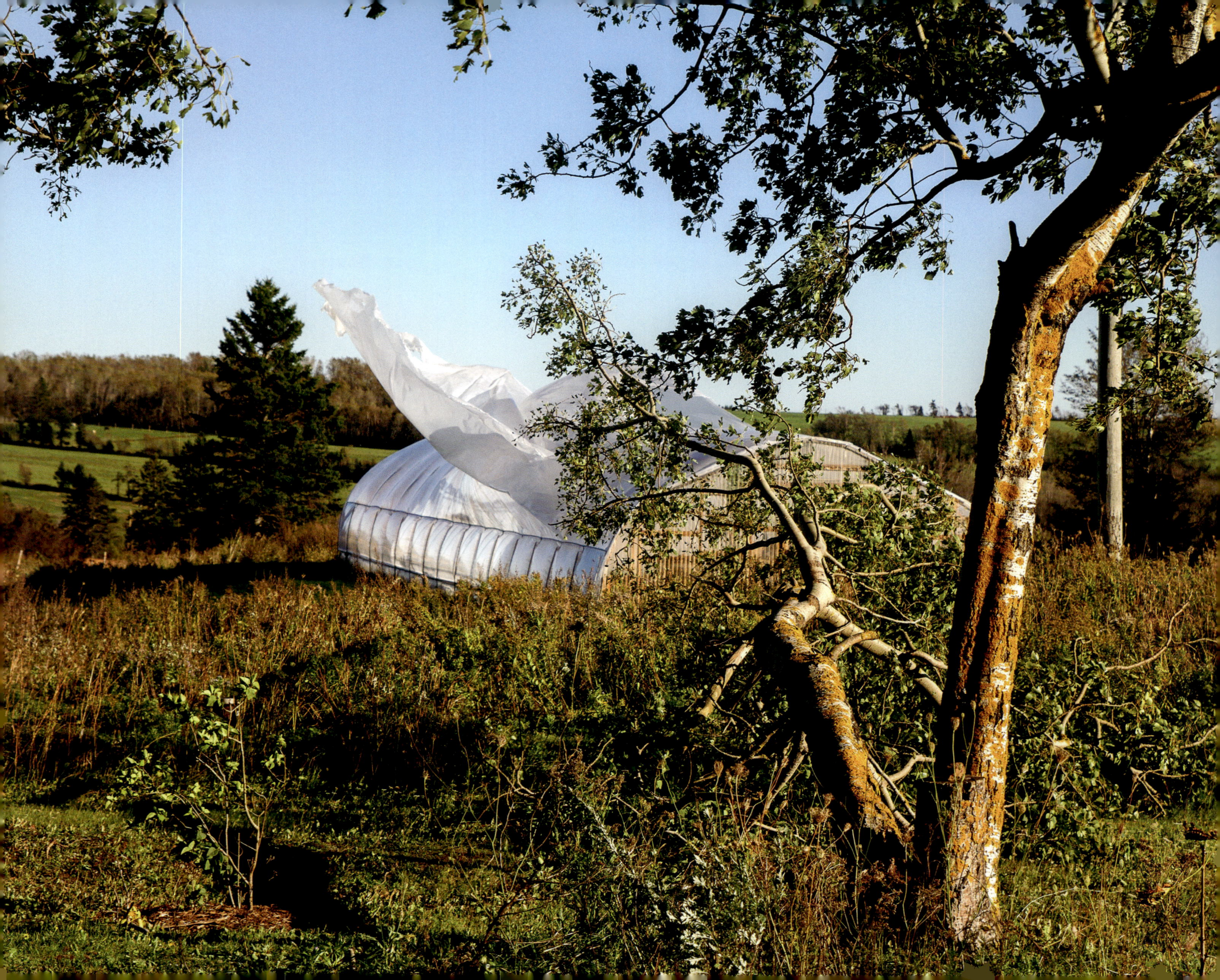

PHIL MERCIER

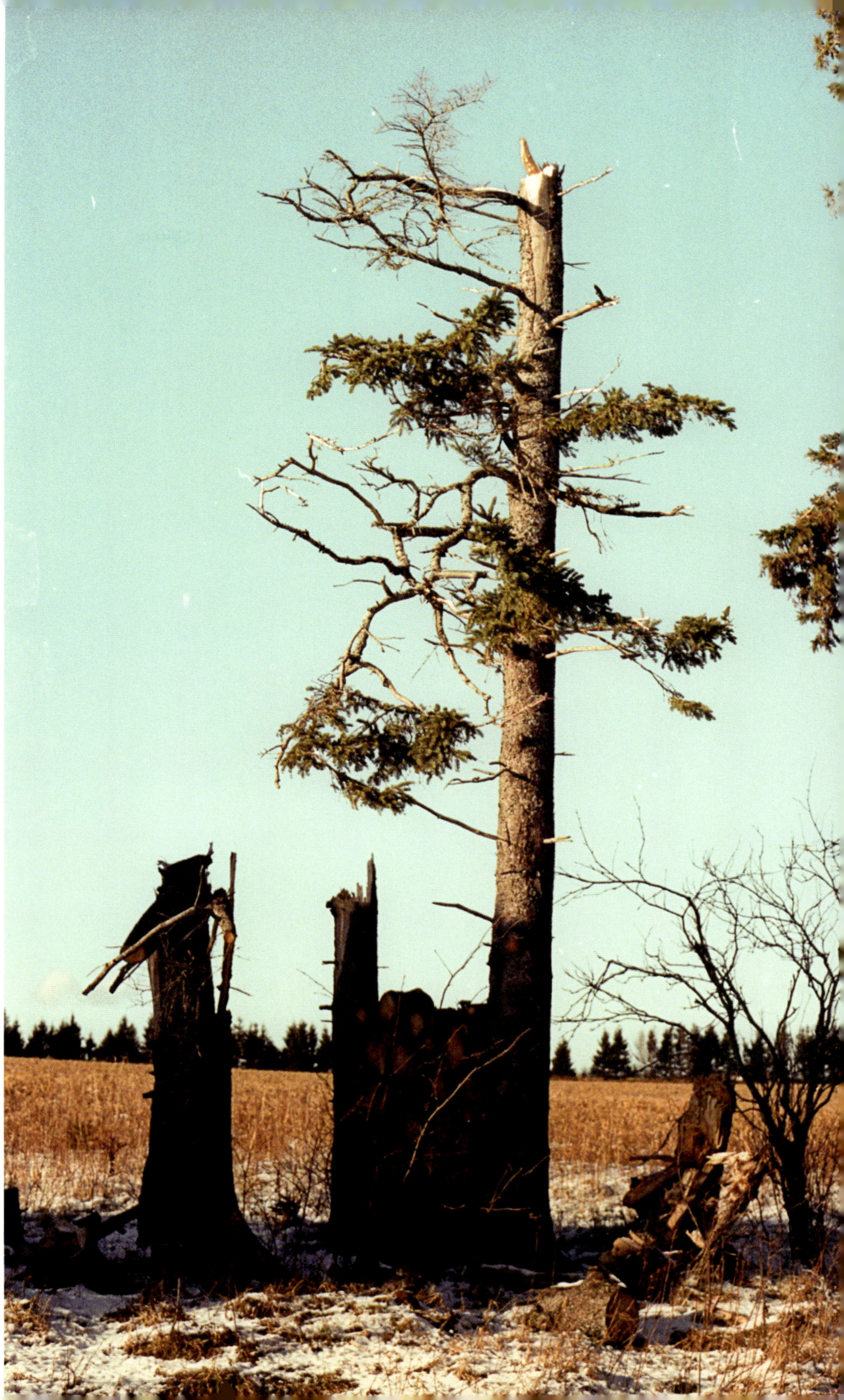

Storm Thoughts

What's left
what left, and what's next

what effect does a hurricane have
on homes, poems, boats, farms
assumed knowns uprooted like all those tree friends of ours
like power rerouted to sky or ground
poof—over
poof—out

what odour is this
of gasoline and ignorance
of propane tanks and hope
we hold chainsaws and branches
we hold our heads in our hands and mope
shake them, nope

we scan the damage, the horizon
eyes widen to comprehend the scope
emotions nosedive, can't cope

with what perspective shall we approach this
do we bring with us our resilience, or a gratitude coach
silver linings are pretty and deceiving, both

but we do need beauty, to get us through the drudgery
the muck of storm cleanup and recovery
amidst ongoing thuggery by those intent on luxury
the othering it takes to take all that they wish
while the rest of the world bears the consequence

it's a wonder we put up with it

that's where deception comes in, doesn't it

we fall for fairy tales
put our faith in human nature and government
corporate greenwashing and economic stimulus
as if any weather system ever listened to false promises
with as much ignorance as we do

of course, as ignorance is bliss, it is chosen too
there is no way to respectfully deny
that there aren't links between us and sky
between the shaking ground and what we extract from it
the flooding waters of the world and what we ask from it
no—demand

if the balance seems to be tipping past our favour
we might task ourselves with the art of recalibration

what's left is our world worth saving
our wants and needs that need differentiating
our human nature that is generous enough to help our neighbours clear
trees and carry water
and greedy enough to need some reigning in
I, too, would like endless resources
I do feel sad that I can't have it

what left, this storm, might grow back if we let it
let's replenish trees but nip greed in the bud
let's grieve losses and lives lost in storms far worse than this one

what's next is uncertain
surely winds will rage and clouds will block our vision
may hindsight and insight guide our transition
to the wise love we need
for all future weather systems.

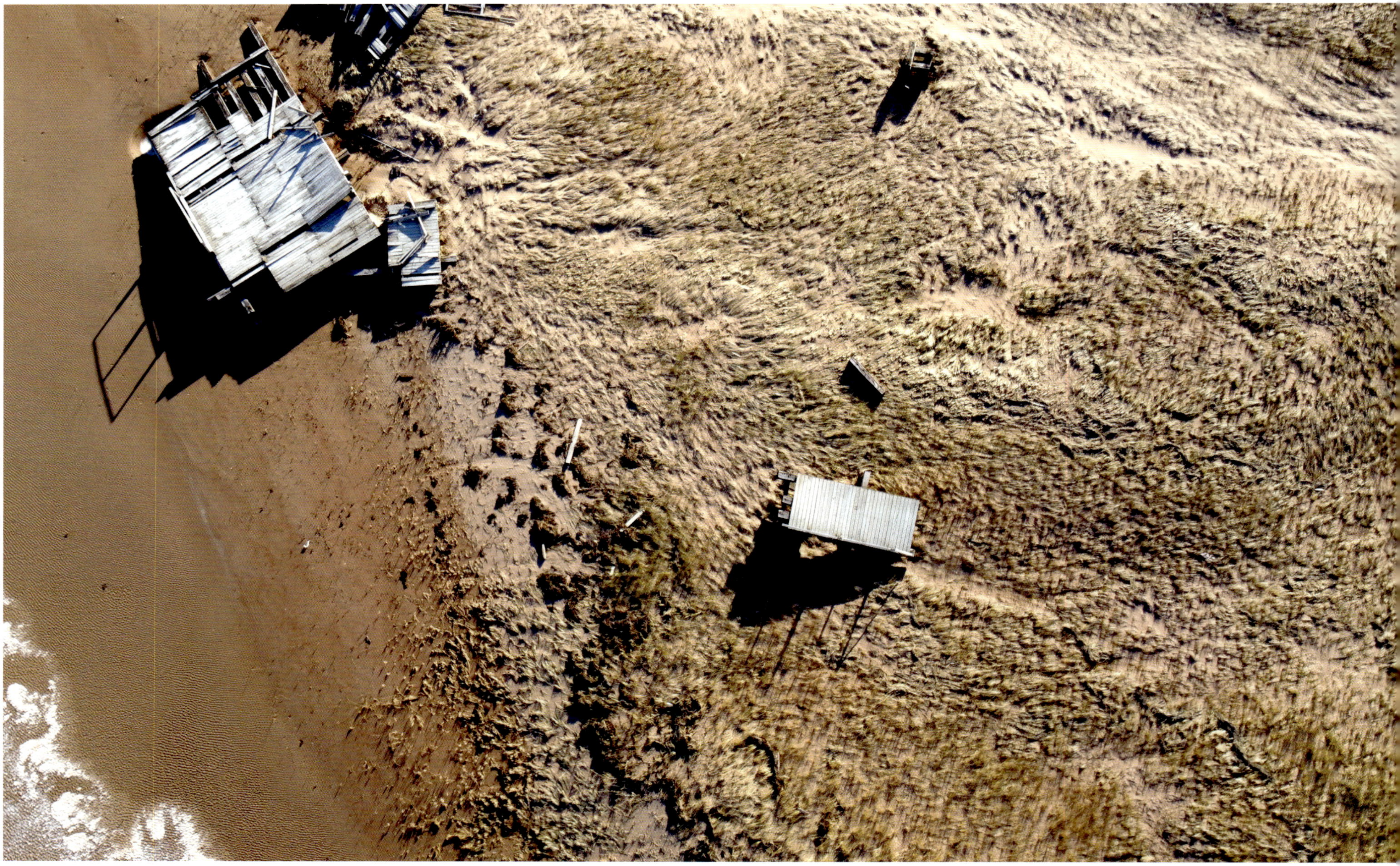

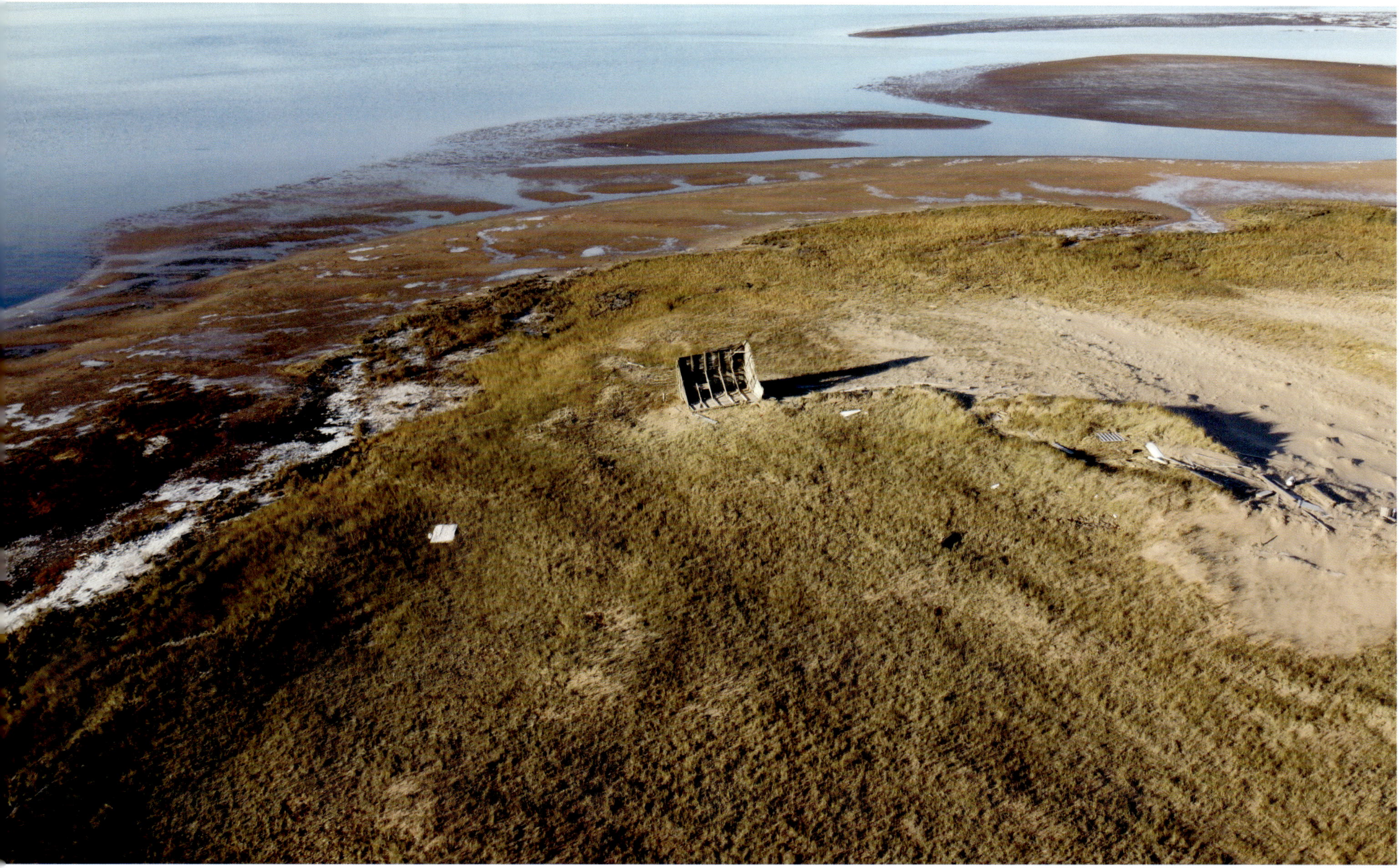

A Message from the Publishers

WHEN POST-TROPICAL CYCLONE FIONA HIT, it felt like the world might just stop turning.

No one on Prince Edward Island was untouched; we all felt the impact. This anthology represents our witness to this most unique weather event. Fiona may have taken our trees, taken our roots, and taken our breath, but she also gave us the pause to grieve, in quiet moments, hand in hand with our families, our neighbours and our friends.

Every day for twenty-one days, Islanders became unburdened from Zoom calls and ringing phones. We leaned towards one another as we leaned towards the light, finding solutions in the rising hum, a sea of generators amongst the churning waters of our bays. Even as we became undone, children laughed again, climbed and played. The pause was brief as, one by one, the coffee makers came back on, the sighs of the neighbourhood audible. It was a team effort, but together, we collected ourselves; we made it.

We stand in a place now where we must look forward, not back. The lessons are there, and this anthology reminds us of those moments, in the depths, when we were touched by fear, by trepidation and by unease. But it also celebrates our connection to one another, to our community, and to the land.

As we brace for the climactic unknown, we know that where we go from here, we will go together. We are certain Prince Edward Islanders will become more and more ready for the storms of tomorrow, and slowly our trees will rise again towards the sky.

—Maureen & Genevieve
Pownal Street Press

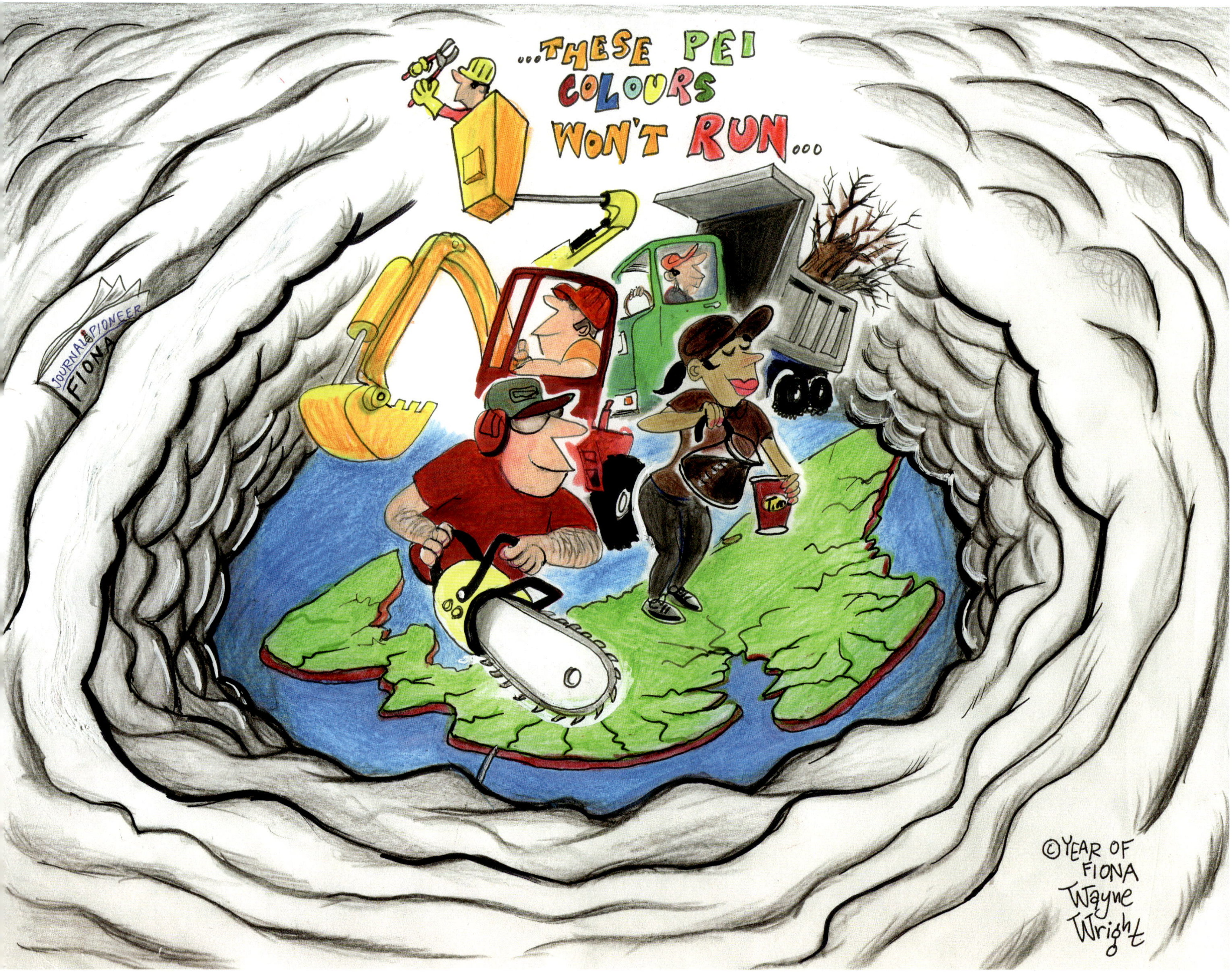
...THESE PEI COLOURS WON'T RUN...
JOURNAL PIONEER
FIONA
©YEAR OF FIONA
Wayne Wright

Acknowledgements

IN THE SPIRIT OF RECONCILIATION, Pownal Street Press acknowledges that the land upon which our organization works is unceded Mi'kmaq territory. Epekwitk (PEI), Mi'kma'ki, is covered by the historic Treaties of Peace and Friendship. We pay our respects to the Indigenous Mi'kmaq People who have occupied this Island for over 12,000 years: past, present and future.

Creating a photographic anthology in under a year is a huge accomplishment, and we would like to thank every contributor that trusted us with your work. Your words, images and art will act as an important record of an event that has forever changed the landscape of Prince Edward Island.

A special thank you to meteorologist Jay Scotland of CBC for your insights before and after the storm. We really appreciate your introduction to the book and your voice shining through on the radio as we waited for the lights to come back on.

Kim Griffin, we appreciate you so much. Thank you for being the Maritime Electric spokesperson, and for standing alongside us as we waited to get the power back. Thank you also for your support with this project. We, as Islanders, will always hold a special place in our hearts for your service and commitment post storm.

Thank you to the Pownal Street Press team: book designer Jordan Beaulieu, editors Mo Duffy Cobb and Sam Wandio, organizer and metadata wizard Ashley Macdonald, and publisher Genevieve Loughlin. This book would not exist without your dedication and commitment to this project.

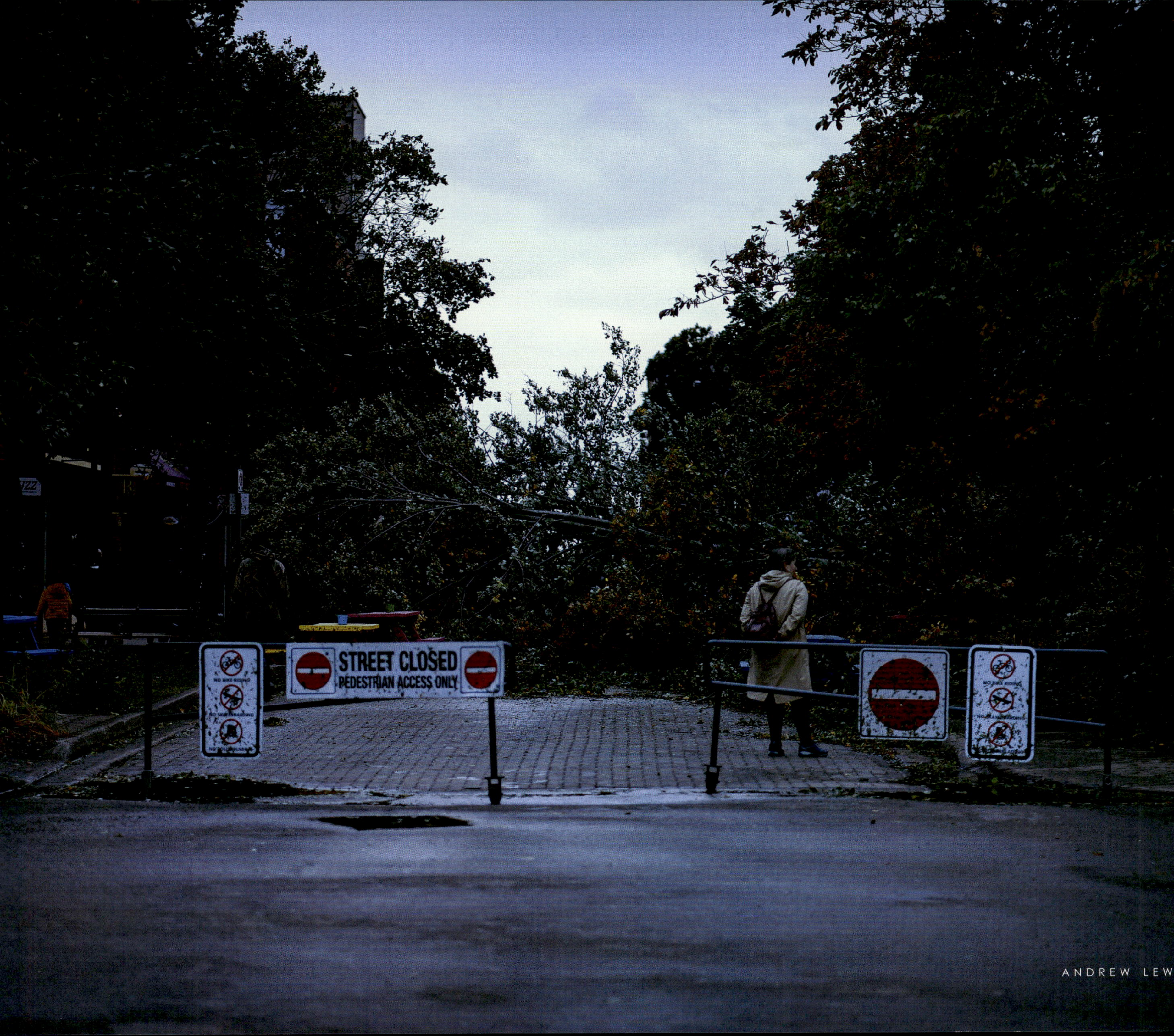

ANDREW LEW

IMAGE CONTRIBUTORS

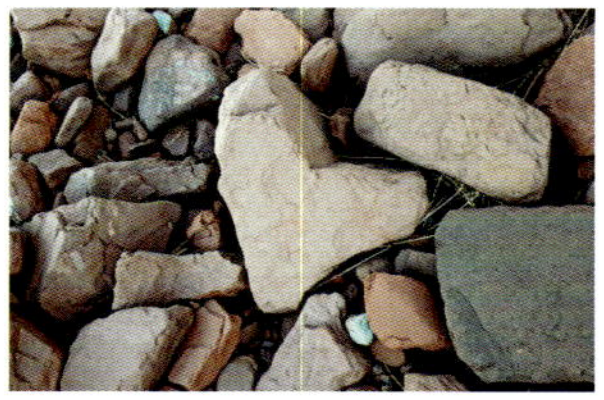

ADRIAN MCNALLY SMITH
99

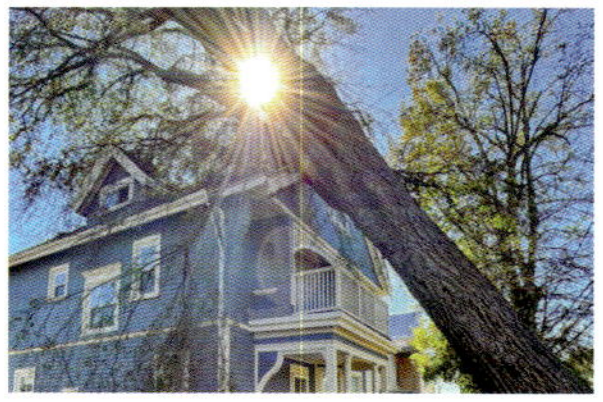

ALANA SPRAGUE
79

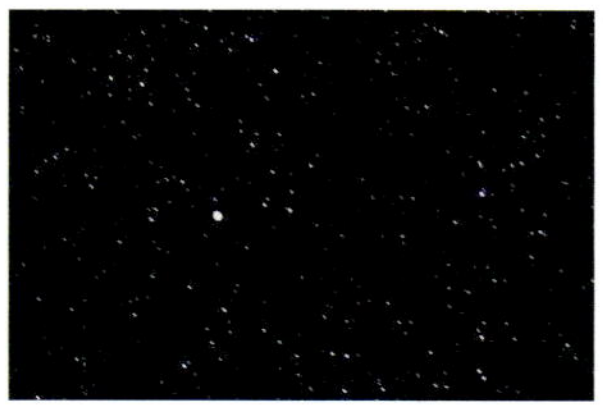

ALEC BEATON
50, 51

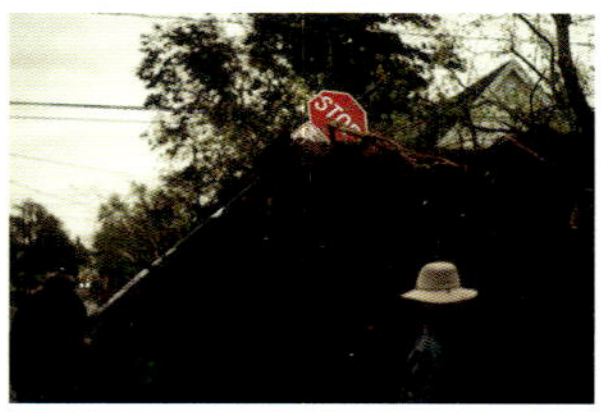

ANDREW LEWIS
Cover, 4, 77, 83,
102, 103, 105, 115

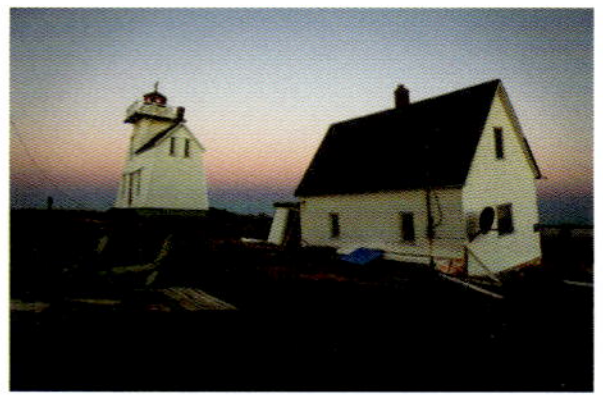

ANGELA ROWLINGS
6, 23, 28, 46, 58, 67,
71, 73, 74, 96

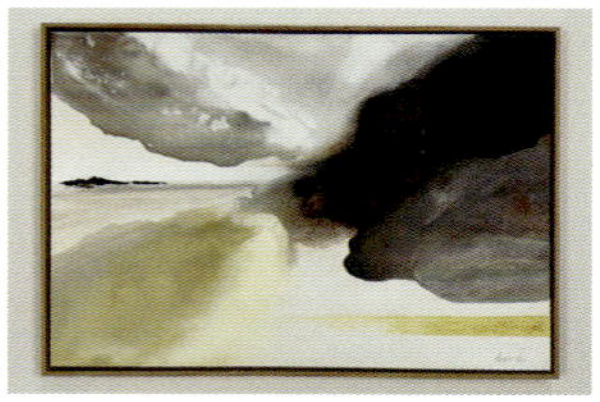

ANGIE ARSENAULT
35

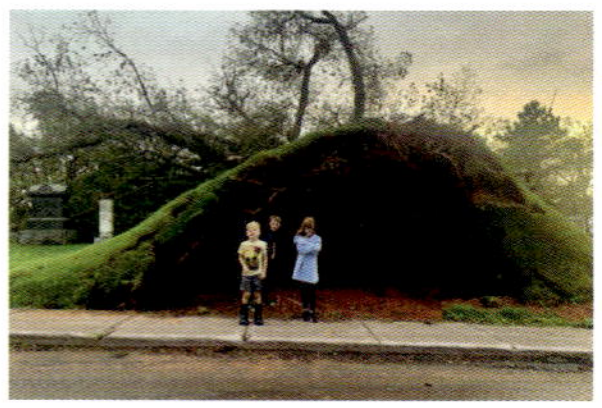

ASHLEY MACDONALD
32

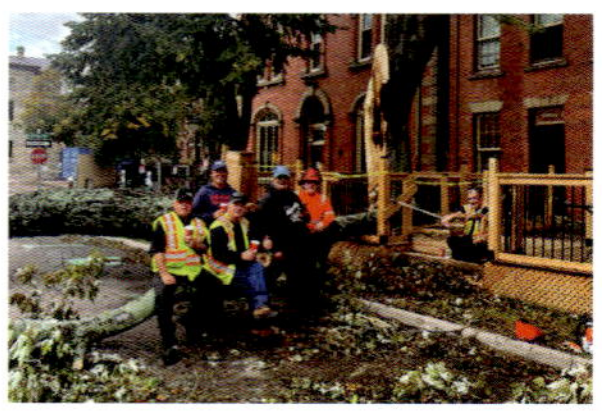

BECCA GRIFFIN
79

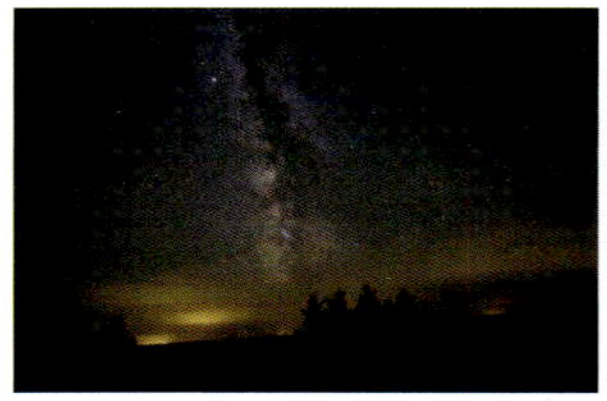

DALE ZAJICEK
25

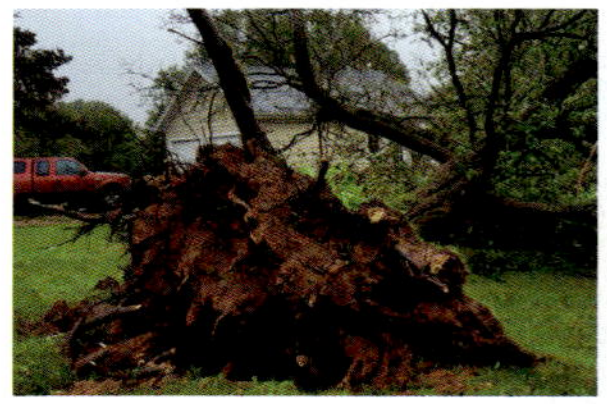

DAWN MACDONALD
93

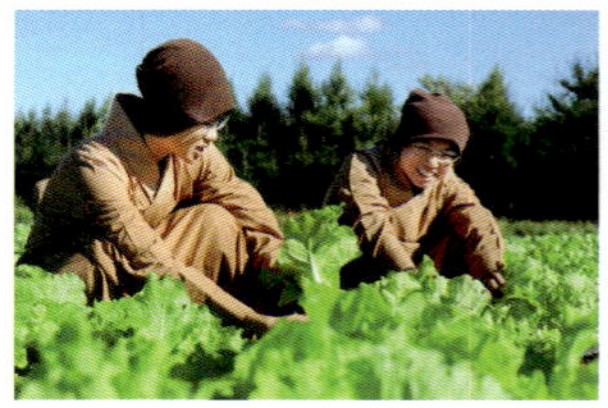

GREAT WISDOM BUDDHIST
INSTITUTE
87, 88, 89

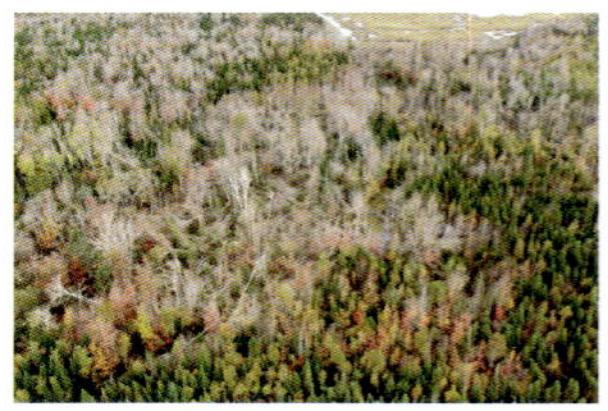

ISLAND NATURE TRUST
8, 30, 31, 57, 81, 90, 92,
93, 110, 111

I M A G E   C O N T R I B U T O R S

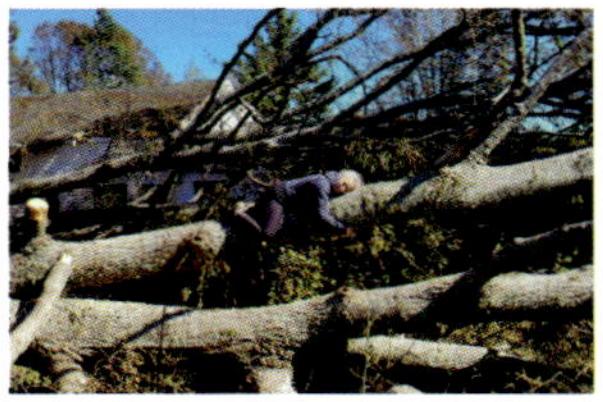

JULAIN MOLNAR
93

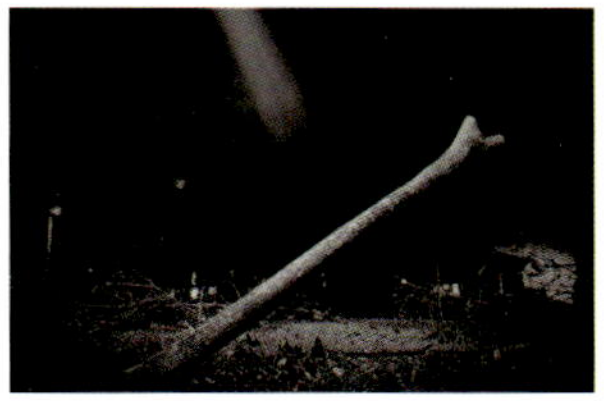

MILLEFIORE CLARKES
62, 63, 65

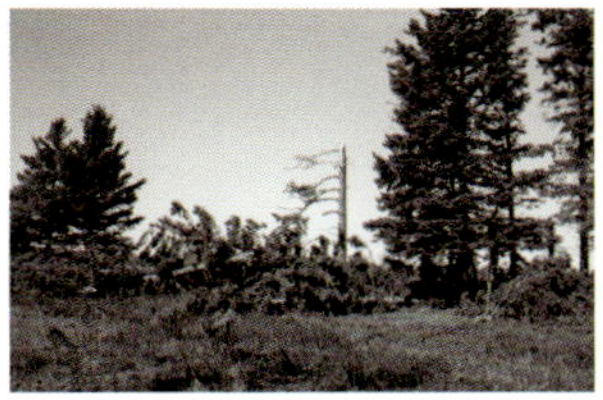

PHIL MERCIER
37, 106, 107

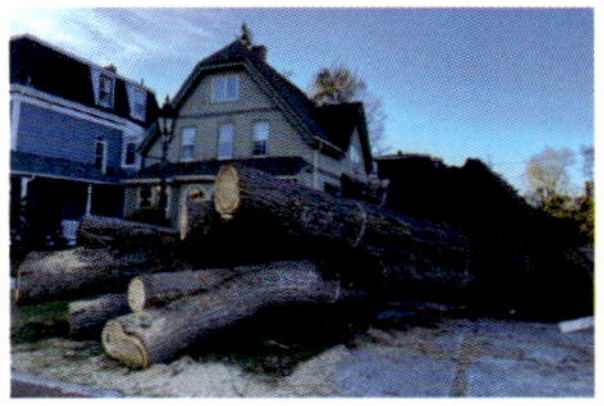

LISA CHANDLER
39

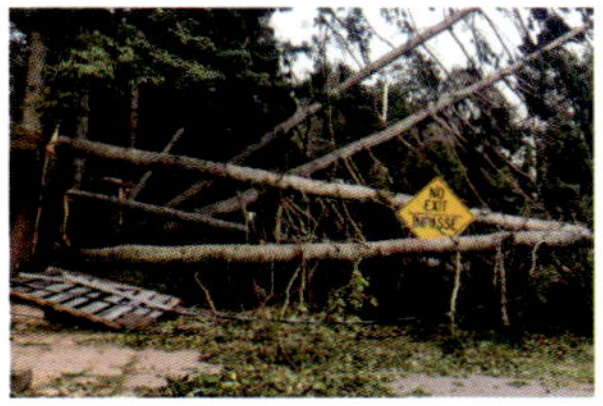

MO DUFFY COBB
21

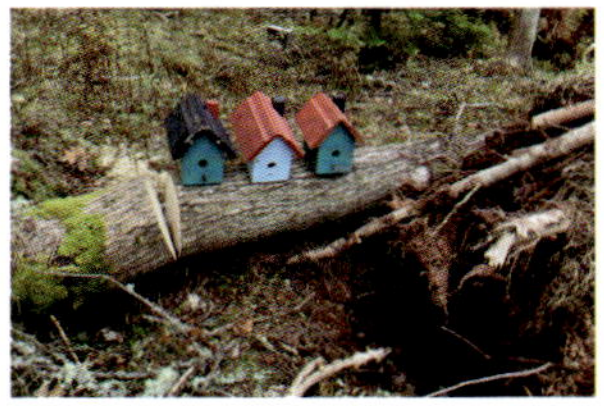

SARA DEVEAU
93

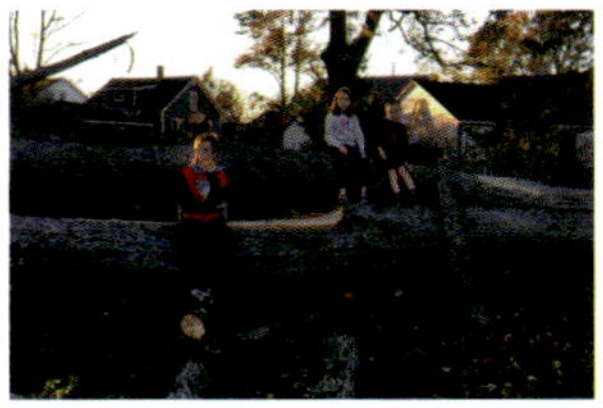

MARILYN CAREY
54

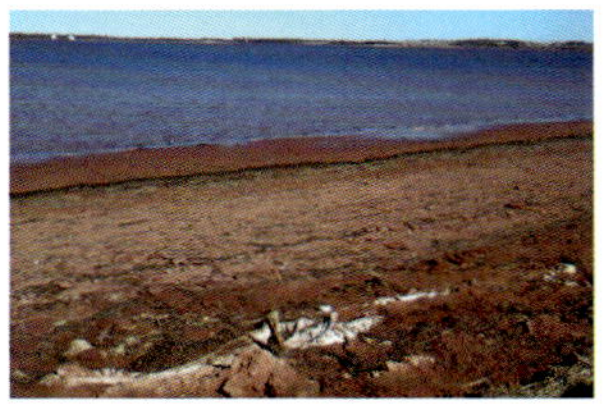

MORGAN MCNEVIN
67

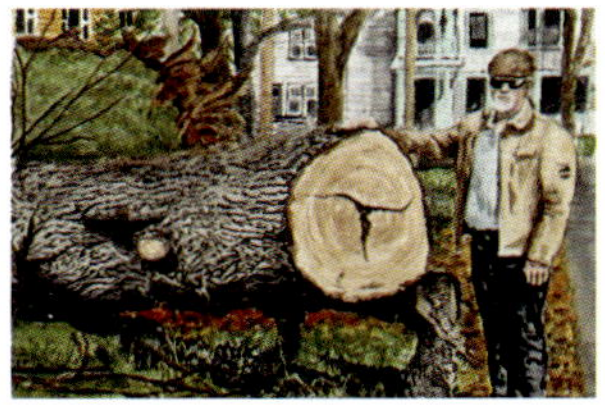

K SHAWN LARSON
44

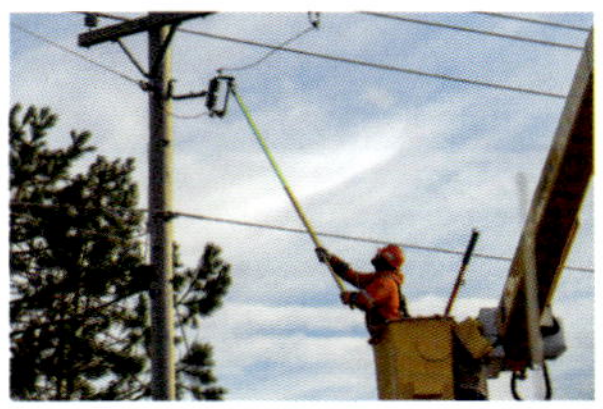

MARITIME ELECTRIC
68, 70, 72

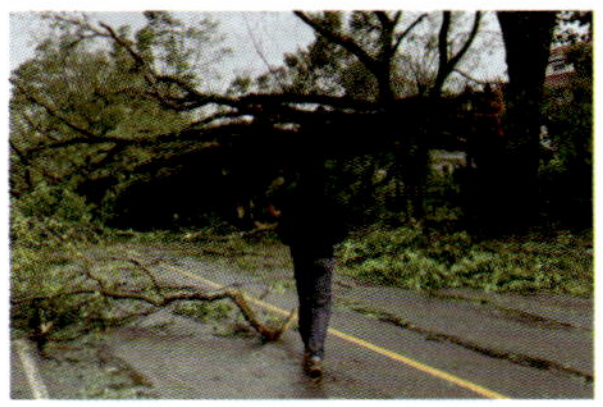

MÉLANIE MORIN
40, 42

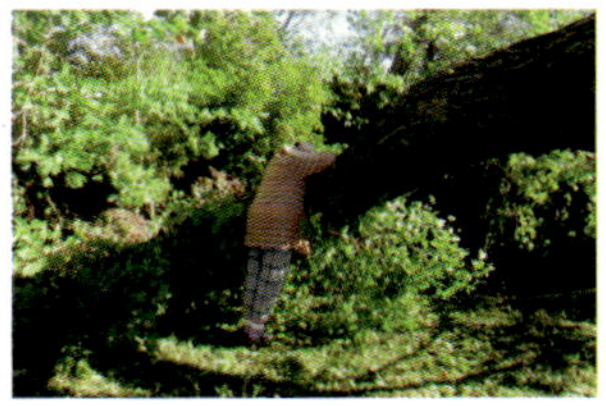

STELLA SHEPARD
17

I M A G E   C O N T R I B U T O R S

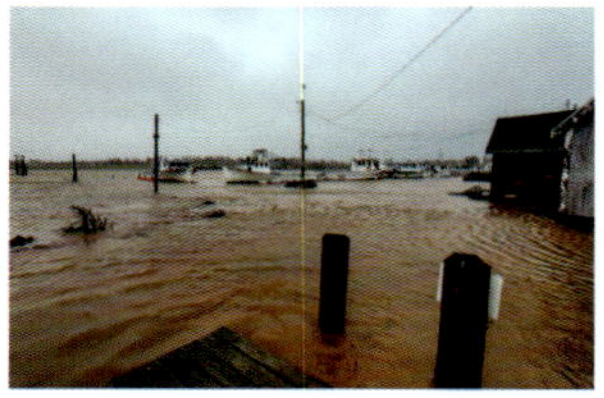

VALFAYE SHAFFER
48, 49

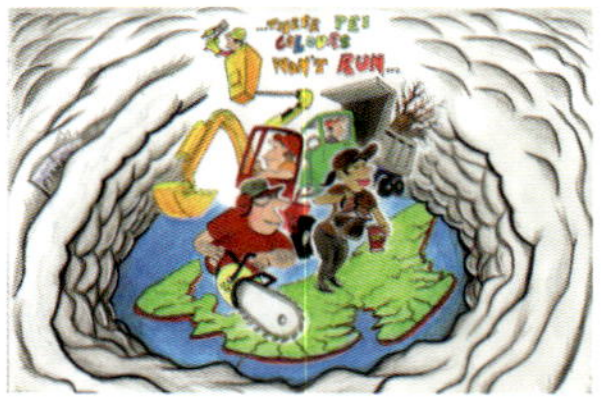

WAYNE WRIGHT
18, 19, 94, 95, 113